Advances in Bioclimatology 1

Managing Editor
G. Stanhill, Bet Dagan

Associate Editors
G. L. Hahn, Nebraska
J. D. Kalma, Canberra
R. S. Loomis, California
F. I. Woodward, Sheffield

Advances
in Bioclimatology ___ 1

With Contributions by
R. L. Desjardins   R. M. Gifford
T. Nilson   E. A. N. Greenwood

With 34 Figures

Springer-Verlag Berlin Heidelberg GmbH

Dr. R. L. DESJARDINS
Centre for Biological and
Land Resource Research
Research Branch
Ottawa, Ontario K1A OC6
Canada

Dr. T. NILSON
Institute of Astrophysics
and Atmospheric Physics
Estonian Academy of Sciences
202444 Tôravere, Tartu
Estonia

Dr. R. M. GIFFORD
Commonwealth Scientific
and Industrial Research
Organization
Division of Plant Industry
G. P. O. Box 1600
Canberra, ACT 2601
Australia

Dr. E. A. N. GREENWOOD
Commonwealth Scientific
and Industrial Research
Organization
Division of Water Resources
Perth Laboratory
Private Bag
Wembley WA 6014
Australia

ISBN 978-3-540-53843-1   ISBN 978-3-642-58136-6 (eBook)
DOI 10.1007/978-3-642-58136-6

Library of Congress Cataloging-in-Publication Data. Advances in bioclimatology. Volume 1 / with contributions by R. L. Desjardins . . . [et al.]. p. cm. Includes index. ISBN 0-387-53843-7 (U. S.) 1. Vegetation and climate. 2. Bioclimatology. I. Desjardins, R. L. II. Series: Advances in bioclimatology: 1. QK754.5A38 1992 581.5′222–dc20 91-26999

© Springer-Verlag Berlin Heidelberg 1992
Originally published by Springer-Verlag Berlin Heidelberg New York in 1992

Typesetting: K + V Fotosatz GmbH, Beerfelden, and Thomson Press India Ltd., New Delhi
31/3145-5 4 3 2 1 0 –

Preface to the Series

Advances in Bioclimatology – the study of the relations between the physical environment and the form and function of living organisms – have been spectacular during the last third of this century.

Before this period, the subject, having slowly emerged from its classical origins as a branch of natural history, had reached the stage of a collection of largely empirical, statistical relationships between standardized but often inappropriate climatological and biological measurements.

Since then, research into the basic physical and physiological mechanisms involved has used the latest techniques of measurement and analysis to develop various bioclimatic relations which have contributed much to improving crop and animal production and optimizing the human environment.

Recently, some of these relationships have been incorporated into larger models of climate-ecosystem interactions. Such models are being used to assess the often unintended effects of human activity on various elements of the biosphere.

However, the advances described have been very unevenly spread through the vast field of interest encompassed by bioclimatology; the fields of plant, animal and human climatology have largely advanced in independent fashions and even within each biological province different techniques of analysis and measurement have developed for different time and space scales of organization.

One of the major aims of this new review series is to overcome this separate development by providing a common forum for those wishing to obtain an authoritative review of the latest developments in bioclimatology.

The emphasis will be on advances which are soundly based on biological and physical principles rather than those describing empirical relationships. Reviews will also deal with the latest techniques of measurement and analysis which are of relevance to bioclimatology and to those describing broader ecological studies in which bioclimatology provides a major element.

Although most of the reviews to be published will be commissioned, the editors would welcome suggestions from individuals interested in contributing a review of the type described, as well as for ideas on major topics of wide interest around which a number of individual reviews could be centered.

Bet Dagan, Israel

G. STANHILL
Editor

List of Editors

Contents

Deforestation, Revegetation, Water Balance and Climate:
An Optimistic Path Through the Plausible, Impracticable
and the Controversial
E. A. N. GREENWOOD

Review of Techniques to Measure CO_2 Flux Densities from Surface and Airborne Sensors

R.L. Desjardins

Contents

1 Introduction

Carbon dioxide is absorbed by vegetation through the photosynthetic process and is released to the atmosphere through respiration of living organisms and decomposition of organic matter. It is also one of the major waste products resulting from the combustion of various fuels. For thousands of years the CO_2 concentration was thought to have been relatively constant. However, since 1860 the amount of CO_2 in the earth's atmosphere has increased from approximately 270 ppm to approximately 340 ppm (Bach 1978), with a present annual rate of increase of about 1.5 ppm (Beardsmore and Pearman 1987). The increasing level of carbon dioxide and other biogenic trace gases in the atmosphere has led to many scientific speculations on their long-term effects, such as changes in global temperature and precipitation.

Reasonably accurate estimates have been made for the input of carbon dioxide to the atmosphere from the combustion of carbon-based fuels (Roether 1980). Much less understood are the effects of climatic change on the exchange rates between the atmosphere and the biosphere (Bolin 1977; Brown and Lugo 1981). Ecosystems need to be characterized with respect to their potential for gas exchange with the atmosphere in order to predict changes in concentration of radiative gases (Mooney et al. 1987).

Micrometeorologists, physiologists and ecologists have attempted for the last 30 years to quantify the CO_2 assimilation by vegetation. Net CO_2 assimilation measurements over a wide range of environmental conditions, for plants in the same physiological state, may provide the required information to sort out the influence of closely correlated elements such as temperature, radiation and wind on crop growth. Early micrometeorological approaches have been based on measuring concentrations of CO_2 at various heights and calculating an exchange coefficient for diffusion. The difficulty of accurately determining exchange coefficients has prevented the widespread application of these techniques. More direct methods such as enclosure techniques have provided useful information on a small scale. The uncertainty about the accuracy and the difficulty of extrapolating such local measurements to regional scales has limited the usefulness of these data to assess the effects of human activities on the biosphere.

The eddy correlation technique, a micrometeorological method not requiring exchange coefficients, has been developed over the last 20 years. For a long time the lack of fast response sensors for CO_2 and vertical wind has delayed its practical application. Suitable sensors are now available, but their costs still limit widespread use. Ground-based eddy flux measurements, integrating exchanges on a scale of tens to hundreds of meters, should be particularly useful to serve as guidelines for selection of optimum management practices on a field scale. However, they still do not provide information on a scale useful to map the sources and sinks of CO_2 and other trace gases over major ecosystems in the biosphere. Measurements using aircraft-mounted sensors, which provide data on the scales of tens to hundreds of kilometers hold the promise to quantify exchange rates relevant to the global balance of CO_2.

It is the purpose of this chapter to review the state of the art in CO_2 flux measurements and to discuss the limitations and advantages of various techniques. Since indirect measurements of CO_2 flux have previously been reviewed (Norman and Hesketh 1980), these methods will be only briefly discussed and emphasis will be given to the eddy correlation technique which provides the most direct measurements. Although many are actively involved in this area of research, as a matter of convenience, most of the examples used to illustrate the basic principles are taken from the author's own research.

2 Surface-Based CO_2 Flux Measurements

Most of the available surface-based methods to measure the air–surface exchange of trace substances have been reviewed in recent publications (Kanemasu et al. 1979; Businger 1986; Hicks 1986). They are primarily based on micrometeorological techniques, net exchange within enclosures, and isotopic methods. Factors such as sampling requirements, size of site, availability of sensors, cost, etc. usually dictate the method selected.

2.1 Aerodynamic Techniques

One of the earliest applications of these techniques consisted in measuring CO_2 gradients between 1 and 22.5 m above a wheat field (Huber 1952). Quantities proportional to the CO_2 flux were obtained by multiplying the CO_2 gradients by the mean horizontal wind speed at an intermediate height. These data showed marked diurnal variations which, as expected, changed direction at sunrise and sunset.

The lack of insight into physical transfer processes such as turbulent transfer was apparent and led to the development of the aerodynamic techniques. These techniques relied on the measurements of wind speed profiles and the log-profile theory for estimating the eddy diffusivity of momentum. It then assumed similar transfer coefficients for carbon dioxide and momentum and has been used to estimate CO_2 flux by many research groups over various kinds of vegetated surfaces (Inoue et al. 1958; Lemon 1960; Monteith and Sceicz 1960; Baumgartner 1969; Lemon and Wright 1969; Saugier 1976; Houghton and Woodwell 1980). Once stability corrections were introduced to account for the disturbance in the wind regime by buoyancy forces (Dyer and Hicks 1970; Pruitt et al. 1973; Francey and Garratt 1981) typical errors were estimated between 10 to 30% (Verma and Rosenberg 1975).

2.2 Energy Balance and Bowen Ratio Techniques

CO_2 fluxes have also frequently been obtained by energy balance consideration. This is basically an indirect way of estimating eddy diffusivity coefficients of water vapor, heat and CO_2 which are assumed to be all equal (Lemon 1967). By interpreting profile data in terms of energy balance, various modifications of the Bowen ratio techniques were developed to estimate CO_2 fluxes (Sinclair et al. 1975; Baldocchi et al. 1981). Again, very realistic values have been obtained with this technique. However, substantial errors can result if energy storage terms in the soil and in the vegetation are not taken into account, if the net radiation is not correctly determined and if the gradients of CO_2, H_2O and temperature are small. CO_2 flux estimates, using gradient techniques for extended periods, have been reported by Biscoe et al. (1975), who successfully compared the carbon assimilation of a barley crop over a whole growing season with its biomass production.

Numerous papers have been written comparing gradient techniques of measuring gaseous exchange (Fuchs and Tanner 1970; Saugier 1976; Sinclair et al. 1975; Verma and Rosenberg 1975). These papers focused on the estimation of measuring errors rather than the theoretical limitations of the basic principle of flux-gradient theory, which assumes that local gradients determine the flux. This assumption is not justified if eddies are large compared to the scale of curvature of the profile and if the quantity transported has sources and sinks in the mixing region, such as is the case for CO_2 transfer within vegetation.

Observation of short-term fluctuations of scalars confirms that turbulent transfer is very intermittent and that scalars are not necessarily transported along their mean concentration gradients (Denmead and Bradley 1987). Such intermittency can be observed in the time traces of CO_2 obtained within and above a maize crop (Fig. 1; Desjardins et al. 1978).

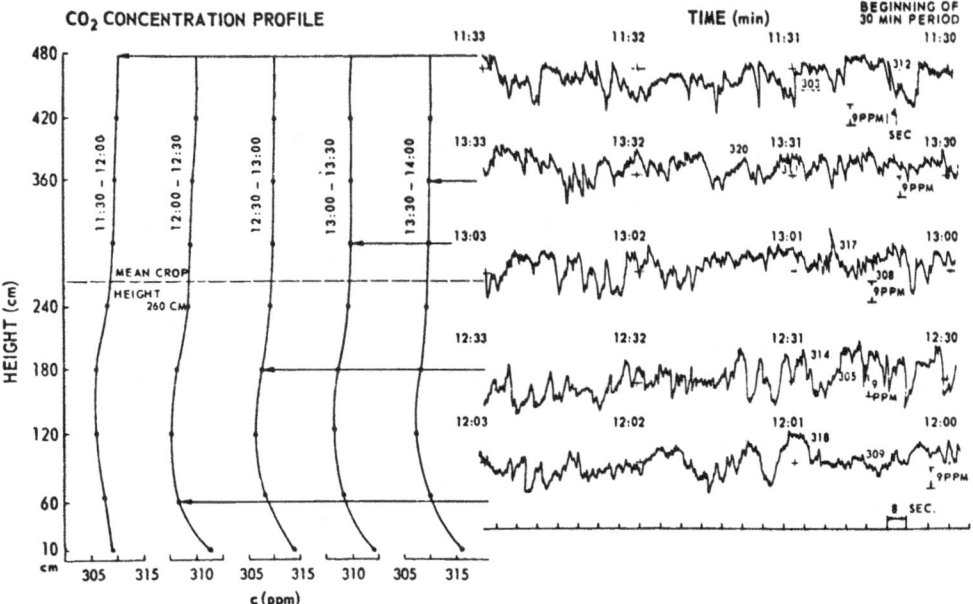

Fig. 1. Short-term CO_2 fluctuations measured simultaneously with mean CO_2 profiles within and above a maize crop

2.3 Eddy Correlation Techniques

The eddy correlation technique involves the measurements of the instantaneous vertical velocity component of the wind, W, simultaneously with a conservative quantity, C. Instantaneous values, C, of any parameter are considered to consist of a mean component, $\bar{C}$, and a transient component, C', with respect to some time scale, so that $C = \bar{C} + C'$. The eddy flux of CO_2, $\overline{WC'}$, that is the covariance of vertical wind and CO_2 is related to the total mean transport, $\overline{WC}$, and the transport due to the mean motion $\overline{W}\bar{C}$ by the following equation:

$$\overline{W'C'} = \overline{WC} - \overline{W}\bar{C}. \tag{1}$$

In practice, $\overline{W}\bar{C}$ is made negligible by rotating the coordinate system of the three wind velocity components such that the vertical and lateral velocity components equal to zero (Tanner and Thurtell 1969). This is necessary because vertical sensor tilt or non-level terrain can cause an apparent mean vertical velocity. In the case of CO_2, Desjardins and Lemon (1974) have found errors up to 10% for 1° tilt.

CO_2 flux data were first obtained using the eddy correlation technique by Desjardins and Thurtell (1970). These data, which were recorded at a height of 4 m above a maize crop (Fig. 2), were overestimated by a few percent because of the lack of correction for density effects arising from the simultaneous transfer of water vapor (Webb et al. 1980), but underestimated by approximately 10% due to the 0.5 s time constant of the CO_2 analyzer (Fig. 4). Density variations can cause large errors in

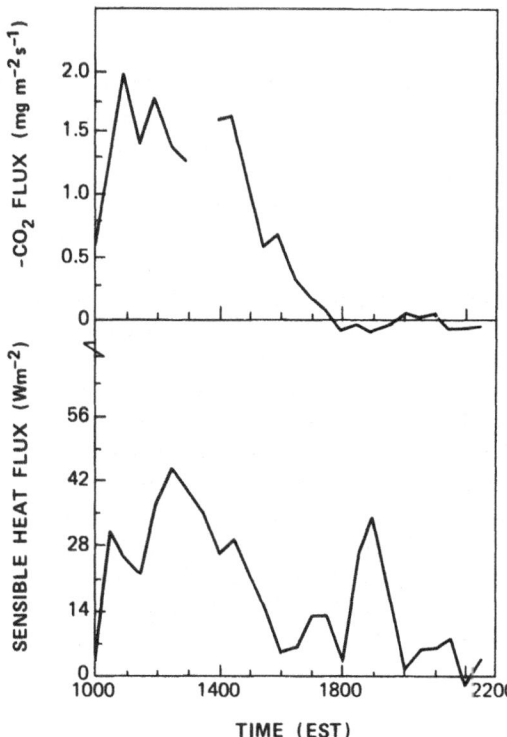

Fig. 2. CO$_2$ and sensible heat flux measured 2 m above a maize crop in 1969

the measurement of CO$_2$ exchange (Leuning et al. 1982), particularly if the sensible heat flux is large such as over dry soil. In that case even the sign of the flux may be wrong.

Methods cannot always be separated into clear-cut categories and hybrids abound. For example, Inoue et al. (1969) developed an assimitron which provided CO$_2$ flux estimates on a 10-min basis. It was the first system to provide the CO$_2$ flux measurements (Fc) directly in the field. It used a sonic anemometer to measure the covariance of vertical and horizontal wind momentum, $\overline{W'U'}$, and the covariance of vertical wind and air termperature, $\overline{W'T'}$. It also required the measurement of the mean vertical differences of density of CO$_2$ $(\overline{\Delta C})$, horizontal wind speed $(\overline{\Delta U})$ and air temperature $(\overline{\Delta T})$. The basic equations were:

$$Fc = \frac{\overline{W'U'}}{\overline{\Delta U}}\,\overline{\Delta C}; \tag{2}$$

$$Fc = \frac{\overline{W'T'}}{\overline{\Delta T}}\,\overline{\Delta C}. \tag{3}$$

Several versions have been developed over the years. In the case of the so-called variance technique, the differences have been replaced by the variances of the scalars (Wesely 1988; Desjardins and MacPherson 1989).

2.3.1 CO_2 Sensors

The availability of a suitable fast response gas analyzer is essential for eddy flux measurements of CO_2. Ohtaki and Seo (1976) and Jones et al. (1978) reported on open-path CO_2 sensors consisting of two infrared beams with two detectors. These instruments had fast response but suffered from electronic noise and instrument drift. More stable open-path infrared sensors were described by Bingham et al. (1978), Brach et al. (1981), Ohtaki and Matsui (1982) and Heikinheimo (1986). They used two detecting bands, only one of which was sensitive to fluctuations of the CO_2, the other being used to monitor instrument stability. Sensors which simultaneosuly measure H_2O fluctuations have now been built by research groups at Okayama University in Japan, Guelph University and Agriculture Canada in Canada. Absorption bands are near 2.6 μm for H_2O and 4.3 μm for CO_2.

The quantity measured by these analyzers is the concentration of CO_2 and H_2O in a fixed volume. The interference of CO_2 on H_2O measurements is usually negligible while the interference of H_2O on CO_2 can be corrected for. An analyzer of this type is shown in Fig. 3 (Chahuneau et al. 1989). It can detect differences of 0.1 mg CO_2 m^{-3} and 0.01 g $H_2O\,m^{-3}$. Its small size (25 cm path length) leads to minimum disturbance of the wind measurements (Wyngaard 1988). Its frequency response of 30 Hz is fully adequate for eddy flux measurements at a couple of meters above vegetation.

Fig. 3. Typical facility presently used to obtain eddy flux measurements of CO_2, H_2O, sensible heat and momentum

2.3.2 Cospectral Estimates

Any meaningful discussion of sensor adequacy for measuring turbulence transfer must be related to the scale of turbulence and turbulence transfer processes as reflected by the cospectral estimates. Cospectra of vertical wind and CO_2 (Cwc) and vertical and horizontal wind (Cwu) observed 4 m above a maize crop in Ottawa, with a fetch of more than 1 km, are presented as an example of the scale of transport for CO_2 and momentum (Fig. 4). Based on these, the contribution of various frequencies to the flux can be calculated (Elagina and Lazarev 1984; Anderson and Verma 1985). For example, with a sensor frequency response of 2.5 Hz, CO_2 fluxes would be underestimated by approximately 10%. The change in sign at high frequencies of Cwu implies a phase shift between W and U signals which can lead to an underestimation of the flux. This is probably attributable to the spacing between the W and U sensors, which was approximately 45 cm.

It is usually advisable to examine the correlation, r, between W and the scalar quantity for various lags and lead times. Frequently, maximum correlation does not occur at time zero and flux calculations should be corrected for that. This is illustrated by an example where the maximum r_{wu} occurred at a lag of 0.1 s, while for the W and CO_2 sensors which were closer together, the maximum r_{wc} occurred at 0.025 s (Fig. 5). In both cases the scalar signal lagged the W signal. This information can be used to evaluate the underestimation of the flux due to spatial separation of sensors and, with proper alignment of sensors in the wind, it can also be used to correct the flux estimates (Chahuneau et al. 1989).

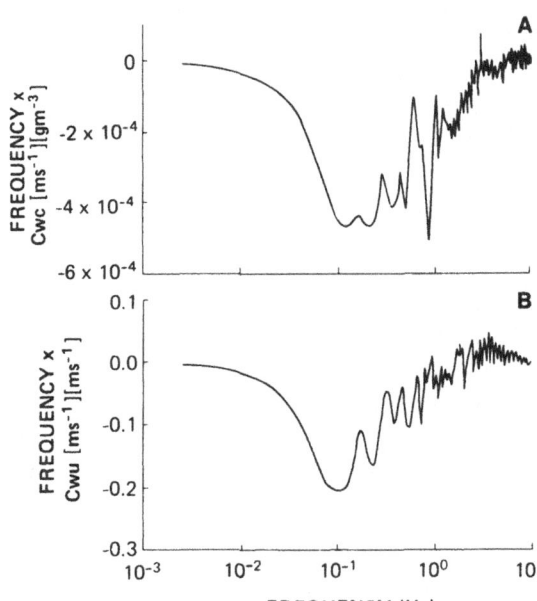

Fig. 4. Cospectral estimates of the vertical wind with CO_2 (**A**) and horizontal wind (**B**) measured 4 m above a maize crop

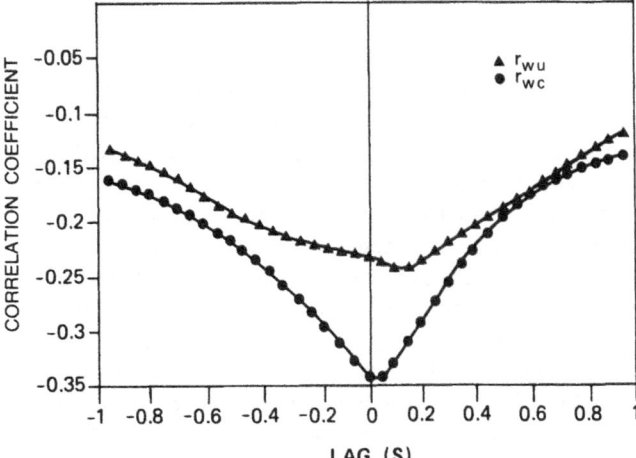

Fig. 5. Cross-correlation of the vertical wind and horizontal wind (r_{wu}) and vertical wind and CO_2 concentration (r_{wc}) with various lag time for sensors mounted 4 m above a maize crop

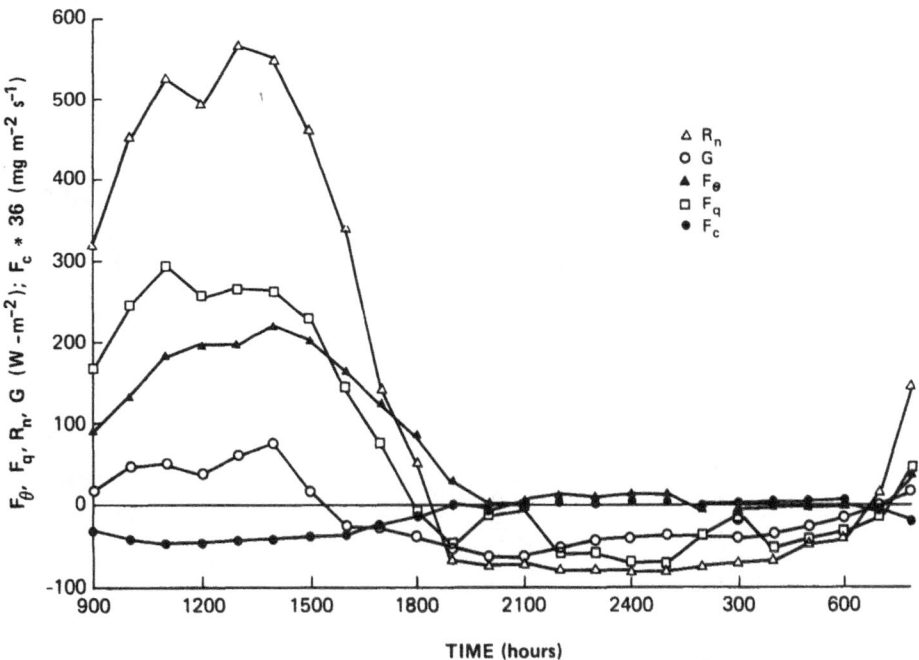

Fig. 6. Flux densities of net radiation (R_n), soil heat (G), sensible (F_θ) and latent heat (F_q) and CO_2 (F_c) measured 4 m above a maize crop

2.3.3 Eddy Flux Measurements

Figure 6 shows an example of the flux measurements now obtainable for extended periods using a ground-based system. Other groups such as Ohtaki (1984), Verma et al. (1986, 1989) and Volkov et al. (1986) have also obtained similar data to characterize crop activity with respect to assimilation, respiration and transpiration. Measurements can be affected by many factors such as poor frequency response of sensors, aerodynamic interference by the tower or sensors, cross-talk between sensors, shadow effect on sonic anemometers, length of averaging period and the degree of non-steady state of the scalar (Ohtaki 1984; Wyngaard and Zhang 1985; Wyngaard 1988; Chahuneau et al. 1989). The effect of large change in concentration over short time periods is particularly important for CO_2 flux measurements. For example, a 11% temporal variation in CO_2 concentration over 30 min, that is approximately 50 mg CO_2 m^{-3}, has been shown to cause a 10% relative error in the flux estimate of CO_2 (Baldocchi et al. 1988). Such variations practically never occur during the daytime but variations of this magnitude, and even considerably larger ones, are common at night under calm conditions. This makes CO_2 flux measurements during these periods completely unreliable. Even with all these potential sources of errors under steady wind conditions, it is now possible to obtain reliable CO_2, latent and sensible flux estimates by using ground-based systems. Energy budget validation shows that eddy flux measurement errors are of the order of 10% (Chahuneau et al. 1989).

2.4 Enclosure Techniques

Enclosures have been used to measure CO_2 exchange of an individual leaf (Field et al. 1982), of all available leaves of a plant (Peterson and Zelitch 1982) and of a number of plants (Saugier 1976; Christy and Porter 1982). Even though they drastically disturb the plant's environment, such techniques provide insight on the physiological response of plants and the variability between plants. Pearson et al. (1936) observed from photosynthetic measurements that top sunlit leaves accounted for 80% of the 3.0 mg m^{-2} s^{-1} canopy flux.

Chambers have also been used to measure the release of CO_2 by soil respiration (Kanemasu et al. 1974; Desjardins 1985; Keller et al. 1986). Such measurements are important because, except for measurement within a forest (Baldocchi et al. 1986), micrometeorological techniques have been unable to separate the soil and vegetation contributions to the CO_2 exchange. As in the case with most chamber techniques, the difference between boundary layer resistance within the enclosure and the atmosphere need to be taken into consideration to form a more realistic basis for extrapolating to regional values. This can cause differences of 100% of the true value.

2.5 Isotopic Techniques

Mass balance measurements over the oceans using radiocarbon tracers and radon 222 have been used to infer exchange rates of CO_2 (Peng et al. 1974; Smithie et al.

1985). Resulting values are 1/10 to 1/20 of most CO_2 fluxes measured over the ocean by the eddy correlation technique (Jones and Smith 1977; Wesely et al. 1982; Greenhut et al. 1983). This has led to considerable debates in the literature (see Broecker et al. 1986). Smith and Jones (1986) have argued that the flux of non-atmospheric gases such as radon does not serve as a complete model for atmospheric constituents such as CO_2. Their contention is that hourly CO_2 fluxes are highly correlated with wind speed and that processes such as wave breaking may accelerate CO_2 exchange. In view of recent improvements in fast response CO_2 sensors and in measuring techniques, more micrometeorological measurements over the ocean are required to resolve the debate and better understand the role of the ocean as a sink of CO_2.

3 Aircraft-Based CO_2 Flux Measurements

3.1 Eddy Correlation Techniques

In order to overcome the implicit spatial limitations of local surface flux measurements, regional observations of fluxes of heat, momentum, water vapor and ozone have been obtained by several research groups using aircraft-mounted sensors (Grossman and Bean 1973; McBean and Paterson 1975; Lenschow et al. 1980; Benech et al. 1987). Since the eddy correlation technique requires observations at only one reference height, it is well suited to measurements from aircraft. Actually, most of the conceptual or practical considerations of tower-based eddy flux observations are applicable to airborne systems. The main assumption that spatial measurements provide information on wind and concentration fields similar to that from fixed-point temporal measurements is certainly justified over an extended homogeneous surface.

Airborne measurement of CO_2 exchanges are of primary interest in two areas of scientific concern: (1) for rapid, large-scale estimates of crop conditions, based on the relationship between net photosynthesis and vegetation indices and (2) for long-term climate studies, concerned with the steady rise in CO_2 content of the atmosphere, where the role of various ecosystems in the global carbon cycle is not yet clearly understood.

Two groups have measured CO_2 flux using airborne systems. The Flight Research Laboratory of the National Research Council of Canada, in collaboration with the Research Branch of Agriculture Canada and MacDonald College of McGill University have cooperated in the development of instrumentation and analysis techniques. They have carried out flights over agricultural crops, forested areas and prairie grassland using a Twin Otter aircraft since 1980 (Desjardins et al. 1982, 1985; Alvo et al. 1984; Schuepp et al. 1989). A U.S. based group involving scientists from Utah State University, Environmental Research Laboratories, NOAA, and Lawrence Livermore National Laboratory have used a NOAA WP-3P research aircraft to measure CO_2 fluxes over the ocean (Greenhut et al. 1983). A schematic of the Twin Otter atmospheric research aircraft used by the Canadian group for geseous exchange measurement is shown in Fig. 7. The sensors and recording system have been described by MacPherson et al. (1987). The aircraft's low speed of approxi-

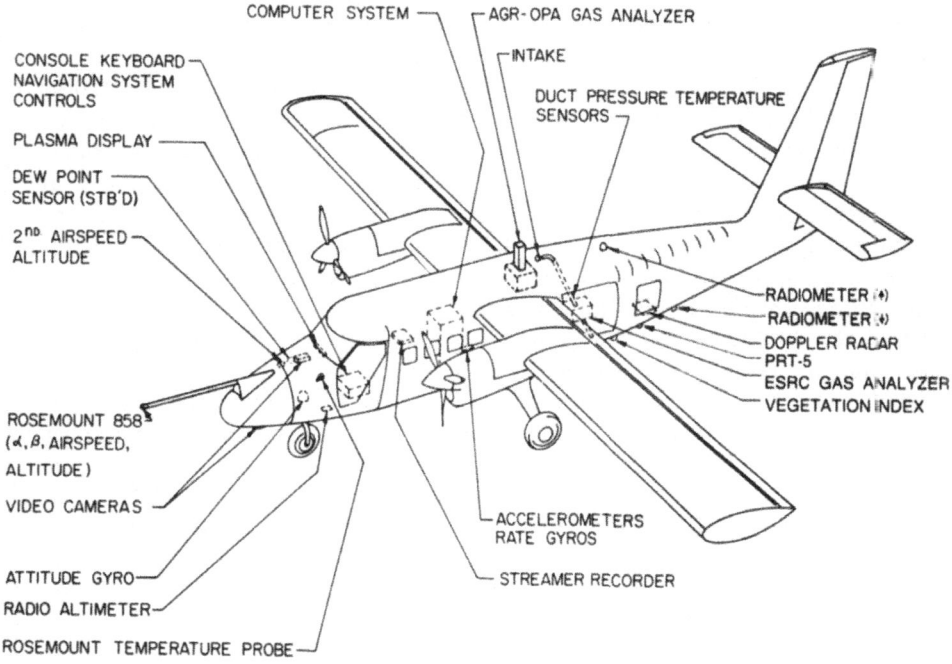

Fig. 7. Schematic of the Canadian atmospheric research aircraft instrumented for CO$_2$ flux measurements

mately $50\,\text{m s}^{-1}$, low altitude capability to less than 10 m, and high maneuverability make it a near-ideal platform for flux measurements. It is instrumented to record the mean and turbulent components of atmospheric motion, along with aircraft position, speed, air temperature, CO$_2$ and H$_2$O fluctuations over a frequency range of 0 to 5 Hz (MacPherson et al. 1981).

CO$_2$ has been measured with an instrument installed either on the roof through the fuselage or within a duct through the rear of the cabin. Air speed, static pressure, air temperature and water vapor fluctuations are also measured close to the analyzer in the duct, to allow computation of the mixing ratio of CO$_2$ with respect to dry air, as well as to compensate in the flux computation for the longitudinal displacement between the nose-mounted gust boom and the analyzer (MacPherson et al. 1987; Desjardins and MacPherson 1989).

3.1.1 Sensor Performance and Sampling Requirements

An important consideration for eddy flux observations by aircraft is the choice of vertical wind and gas sensors. Measurements of the variance of the vertical wind, although relatively difficult to obtain, have been shown to agree within a few percent between Doppler or Litton instruments (J.I. MacPherson, pers. commun.). Such results are very encouraging but more tests are required to fully evaluate the accuracy of turbulence measurements by aircraft mounted sensors. For airborne CO_2 measurements, the principal factors for evaluating the suitability of the sensors are mechanical and electronic stability and frequency responses.

Figure 8 shows typical time series of fluctuations of CO_2 as well as vertical wind speed recorded 25 m above a wheat growing area in Manitoba, Canada. The simultaneous sensible and latent heat flux traces show the correspondence of transfer to the various scalars. Even though the signal from the AGR-OPA contains 30 times more random noise as compared with the ESRC analyzer, some of it associated with the vibration due to the roof installation, the 5-s running mean CO_2 fluxes from both sensors are remarkably similar. This is the case because CO_2 fluctuations are considerably larger than the noise and the vertical wind fluctuations are not correlated with noise from the CO_2 sensors. Turbulence transfer, which is highly intermittent even over homogeneous vegetation (Fig. 8), is very dependent on the number of observations included in the calculations. The required number can be estimated from the integral scale, that is the distance over which, on the average, a variable maintains some degree of correlation. For example, Lenschow and Stankov (1986) have reported that a measurement length of 100 to 500 times the boundary-layer height is required to measure a scalar flux with 10% accuracy. At an altitude of 50 m Schuepp et al. (1989) have calculated that the sampling variability

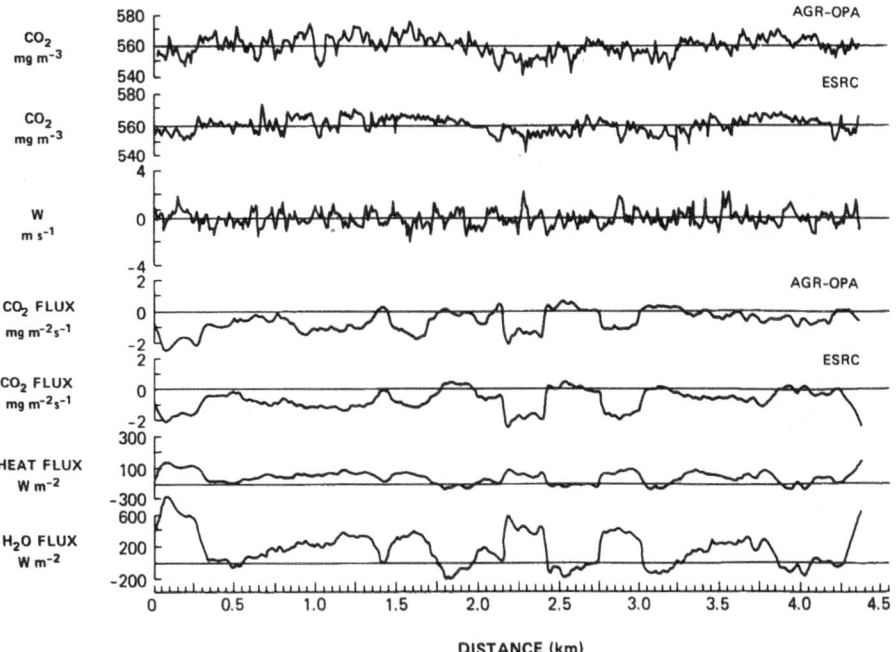

Fig. 8. Time traces of CO_2 fluctuations measured with fast (AGR-OPA) and slow (ESRC) responding infrared analyzers and vertical wind speed for one aircraft pass 25 m above a wheat crop in Manitoba, Canada. The CO_2 fluxes using the two analyzers and the sensible and latent heat fluxes are shown with a 5-s window centered on the plot point

is reduced by approximately 37% when going from 4 to 10 km sampling runs and by about 50% when going from 10 to 40 km. The sampling distance requirement decreases markedly as we get closer to the ground surface and can also be reduced considerably by using area-averaging techniques (Wyngaard 1986).

3.1.2 Aircraft-Based Estimates

Evaluation of aircraft-based flux measurements by ground-based measurements are difficult to carry out because of the very different footprints involved (Saïd 1988). Good agreement was obtained for CO$_2$ flux over a homogeneous wheat growing area 4 × 4 km (Schuepp et al. 1987; Desjardins et al. 1989b). Differences observed

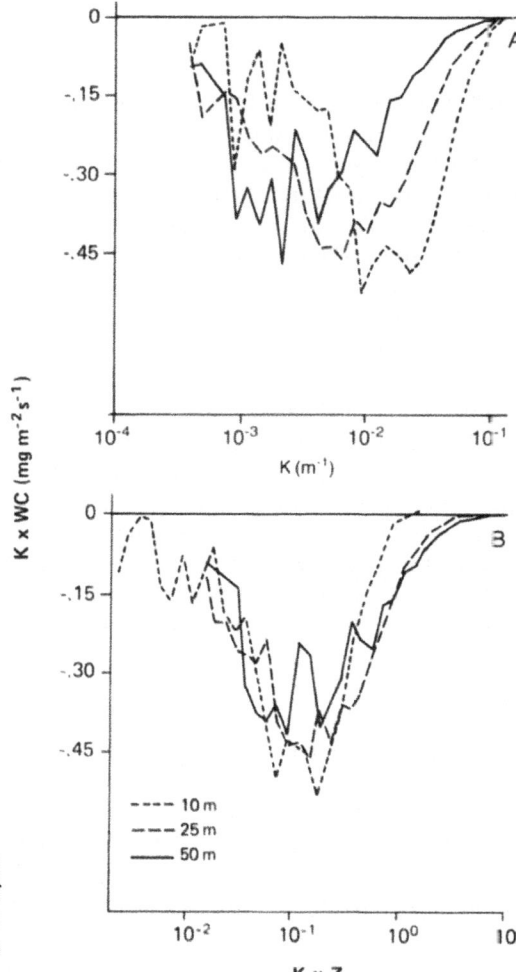

Fig. 9. Average cospectra of vertical wind and CO$_2$ (WC) as a function of wave number (**A**) and non-dimensional frequency (**B**) for three altitudes above a wheat growing area in Manitoba, Canada

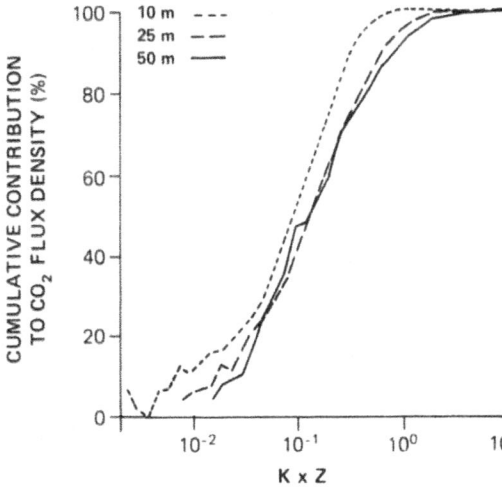

Fig. 10. Average cumulative contribution to CO_2 flux as a function of non-dimensional frequency for three altitudes

could easily be attributed to the long wavelength contributions which were not adequately measured. Mean cospectral estimates (WC) obtained for 77 runs at 10, 25 and 50 m over the area around midday under mainly sunny conditions are presented in Fig. 9A, B. These averages are based on runs obtained during 6 days in July 1985 (Desjardins et al. 1989b). The expected shift to lower wave number, K, of the peak of the cospectra with height due to the larger scale of transfer at higher altitudes is observed in Fig. 9A. In Fig. 9B one notes that once the sampling height, Z, is taken into account through a normalized frequency, $K \times Z$, the cospectral estimates at 25 and 50 m are very similar. At 10 m, an underestimation of appoximately 10% is observed because of the loss of the high frequency contributions of the small eddies.

The cumulative contribution to CO_2 fluxes based on these cospectra is given in Fig. 10 as a function of $K \times Z$. It confirms the similarity in the transfer at 25 and 50 m. These curves are useful for estimating the loss due to various averaging schemes or due to short runs. For example, in the case of the 5-s running mean fluxes presented in Fig. 8, which were obtained at an altitude of 25 m, one can estimate the loss due to a measuring distance of 250 m, that is $K \times Z \approx 0.1$, to be approximately 35% of the value which would have been obtained by averaging the fluxes over a 4 km distance.

3.1.3 Regional Estimates

Flight patterns may vary according to scientific objectives. Some basic ones are shown in Fig. 11. "L" patterns at various altitudes have been useful for extrapolating airborne flux estimates to the surface for relatively large areas. The spatial variability has been explored by grid flights at constant pressure altitude. Profiling runs, 11 km apart, have been used to quantify advection and flux divergence (Desjardins et al. 1988). Runs over long distances have also been done to quantify the long wavelength

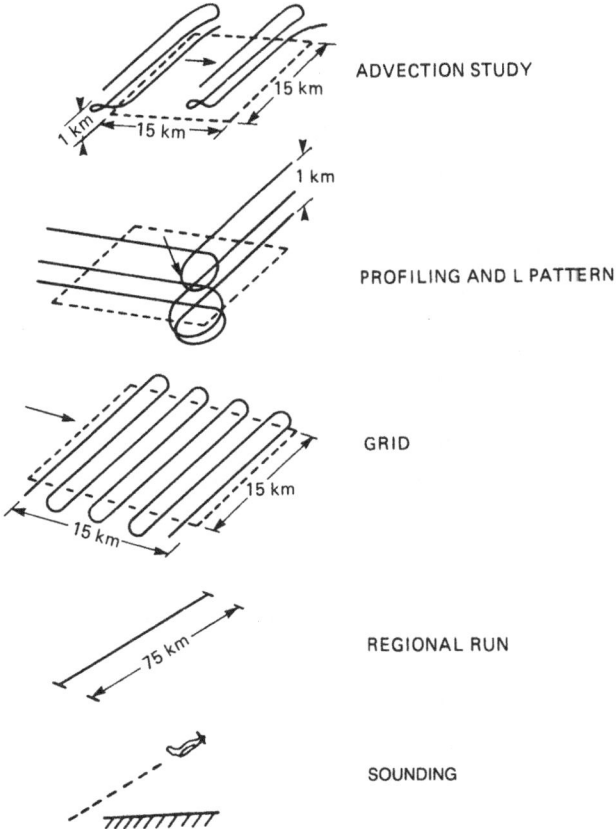

Fig. 11. Flight patterns used to obtain regional flux measurements

contributions of the cospectra and to develop satellite-based algorithms (Desjarcins et al. 1989b; Mack et al. 1989). Most of these data have not yet been fully analyzed but examples of the data obtained during some of these flight patterns will be presented here.

Figure 12 presents CO$_2$, H$_2$O, sensible heat fluxes as well as solar radiation obtained from 16 4 km runs over wheat fields in Manitoba at altitudes of 25, 50 and 100 m during a 1.5-h period. As indicated by the plot of solar radiation, the sky was initially overcast but clearing occurred near 1400 h. The plot illustrates the rapid response of the crop to radiation changes and consistency of flux values estimated over a distance as short as 4 km. These results indicate that such measurements might help to characterize the photosynthetic light response curve of large ecosystems.

In an attempt to map the spatial variations of CO$_2$ fluxes, grid flights have been carried out over a 15 × 15 km grassland ecosystem. These flights consisted of eight grid lines, 1.8 km apart, in an east–west direction under completely sunny conditions around midday at a mean altitude of 80 m with crosswind conditions.

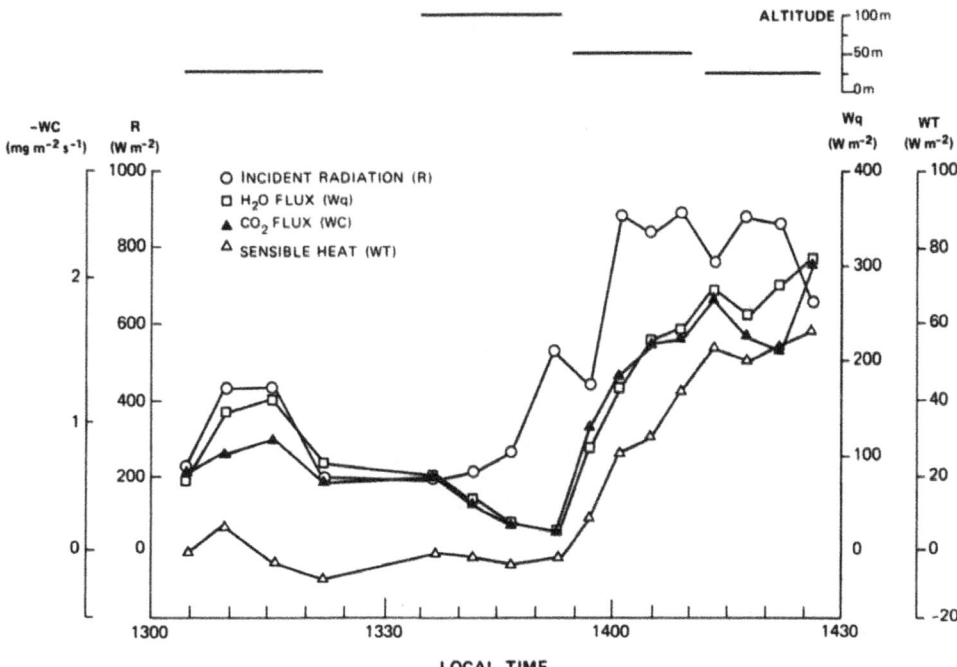

Fig. 12. Run averaged fluxes of CO_2, H_2O, sensible heat and incident radiations obtained over a wheat growing area in Manitoba during a 1.5-h period on July 14, 1986

Three-dimensional plots using a 16×16 bivariate spline (Meinguet 1979) of the CO_2 flux data on July 11, August 15 and October 7 are presented in Fig. 13. These graphs are based on averages over 5.4 km distances. They show the spatial variation in CO_2 flux. No effort has been made yet to compare the variations in CO_2 flux to variations in the vegetation below. These dates, however, correspond to substantial changes in the greenness of the vegetation. Vegetation indices (IR/R) (Tucker 1979), as measured by a downward-looking sensor mounted on the flux aircraft, were 2.7 for July 11, 2.2 for August 15 and 1.6 for October 7 (Desjardins et al. 1989a).

3.2 Dissipation Methods

An approach referred to as the dissipation method is sometimes suggested as an alternative approach to the eddy correlation technique. The fluxes are not determined directly but are derived from turbulence measurements through different statistical quantities (Brutsaert 1984). This technique is based on the rates, ε_c, at which molecular diffusion dissipates the scalar of interest. The spectral density method is most commonly used to estimate the various dissipation rates (Ohtaki 1985). The flux,

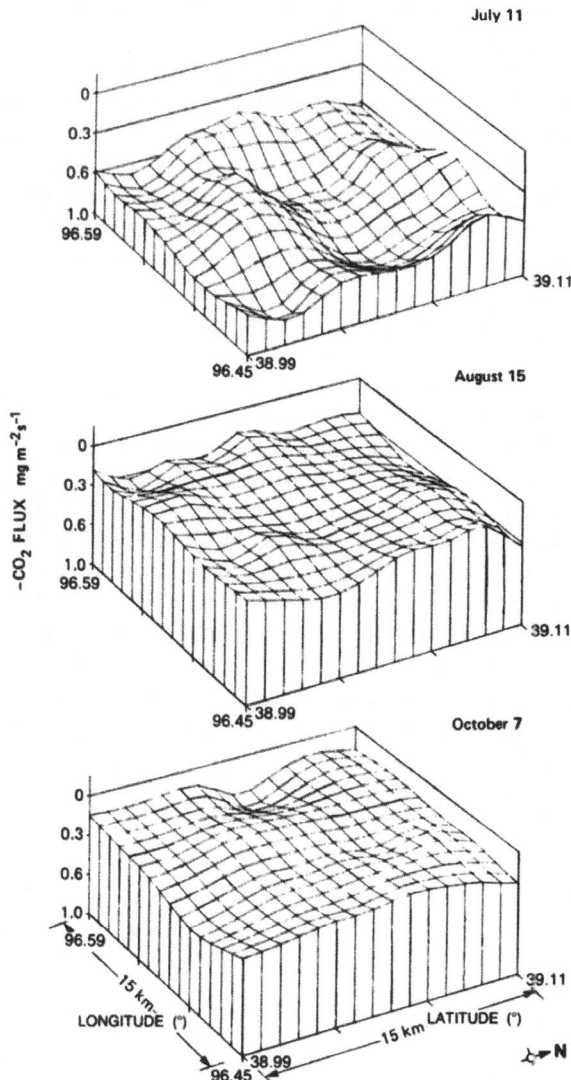

Fig. 13. Spatial variations of CO$_2$ flux over a 15 × 15 km grassland ecosystem on July 11, August 15 and October 7, 1987

F_C, of a scalar is given by:

$$F_C = [k(Z - d_0)u_* \varepsilon_C/\phi_{SC}]^{1/2}, \qquad (4)$$

where k is the von Karman constant, $Z - d_0$ is the height above the virtual surface level, u_* is the friction velocity and ϕ_{SC} is a stability function. Unlike the eddy correlation technique, this method requires no precise orientation and alignment of the vertical wind sensors. The method is still in a developmental stage but it may eventually be a useful tool in cases where the eddy correlation technique might give rise to problems, such as under light wind conditions and complex topography (Brutsaert 1984).

4 Applications, Suggested Further Work, and Summary

Agriculture has been practiced for thousands of years and as a whole, its progress can be viewed as a natural evolutionary selection process: successful agricultural practices evolve as a response to environmental fluctuations that are predominantly weather-induced. During the last century, the emphasis has gradually shifted towards management practices aiming at the highest possible yield.

Yields of many crops have doubled and even tripled due to technological advances in the areas of tillage and planting equipment, fertilizers, pesticides and cultivars. However, along with these advances have come challenges to protect our environment and conserve our natural resources including water, soil and fuel energy.

The increasing number of options available to the farming community makes it more and more difficult to select the most appropriate management practices. The successful choice must consider the whole complexity of the soil–plant–atmosphere interactions. More specific information on photosynthetic and water use efficiencies are required on a short time scale to develop guidelines to arrive at the potential efficiencies estimated by Lemon (1969).

CO_2 and H_2O flux measurements can be used to quantify, on an hourly basis, crop response to environmental conditions for a whole field. However, many experimental tests in agriculture involve either small plots or individual plants. The difficulty in relating leaf measurements to canopy measurements complicates the interpretation of such experiments. The mean hourly CO_2 assimilation measured by an eddy correlation method 2 m above a maize crop is compared in Fig. 14 with

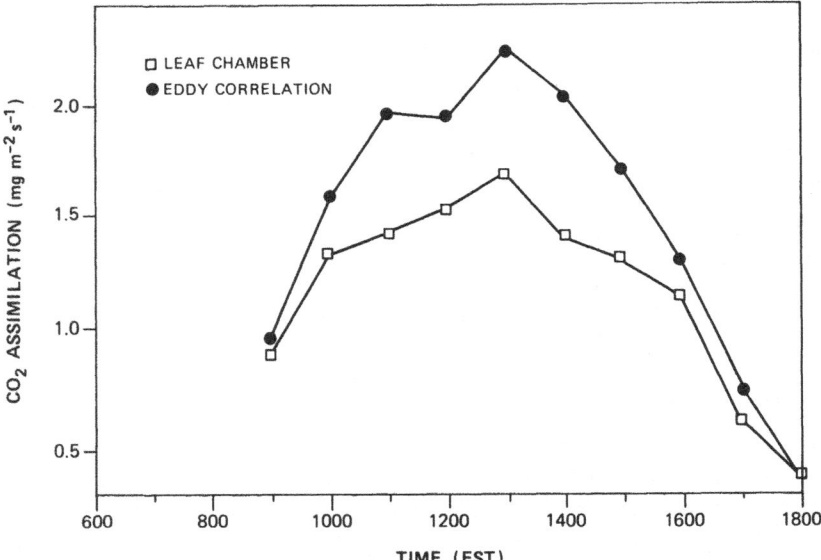

Fig. 14. CO_2 assimilation based on leaf chamber and eddy flux measurements over 1 day for a maize crop in August 1987

the mean hourly CO$_2$ assimilation obtained by averaging ten leaf chamber measurements: four upper leaves, two sunny and two shaded middle leaves and two lower leaves (F.K. Karanja, pers. commun.). As expected, the eddy correlation values, which include the whole canopy, are larger than the ten leaf averages. Such a combination of measurements provide the required data to test methods of estimating net photosynthesis from leaf chamber measurements and crop models. Much progress has been achieved in developing crop-weather models (Shawcroft et al. 1974; Sinclair et al. 1976; Goudriaan 1977; Norman 1979; Stewart and Dwyer 1986), but the testing and validation have often been very limited.

Aircraft-based flux measurements have the potential to integrate exchanges of CO$_2$ on a scale appropriate for imaging the rate of biomass production. Austin et al. (1987) estimated that eight passes were needed to delineate 1 km scale surface features over a hail-damaged area. The study of Mack et al. (1989), which compared CO$_2$ flux measurements to Landsat MSS data for 14 regions in Manitoba, Canada, demonstrated that vegetation and soil types had to be taken into account to obtain good correlations between CO$_2$ flux and vegetation indices, IR/R.

The use of ground-based and airborne CO$_2$ flux data as a mean of calibrating satellite observations is just starting to be exploited. It is the basis of the First International Satellite Land Surface Climatology Field Experiment whose objective is to better understand the role of biology in controlling the interactions between the atmosphere and vegetated land surfaces and to investigate the use of satellite observations to infer climatologically significant land-surface parameters (Sellers et al. 1988). Such large-scale projects are required over major ecosystems if we are to be able to predict their response to changing environmental conditions. Finally, techniques developed for measuring CO$_2$ exchange will undoubtedly provide valuable information for estimating biogenic trace gases such as oxides of nitrogen and methane which also contribute to the greenhouse effect (Mooney et at. 1987) and for which fast response sensors will soon be available.

Acknowledgements. I am grateful to my colleagues P. Alvo, E.J. Brach, D. Buckley, F. Karanja, A. Mack, J.I. MacPherson, E. Pattey, P.H. Schuepp, R. Verdon, P. Poirier and M. Moscos with whom I have worked in developing these techniques during the past few years and who have greatly contributed to the data collection and interpretation of ideas presented in this paper.

References

Allen LH Jr, Hanks RJ, Aase JK, Gardner HR (1974) Carbon dioxide uptake by wide row grain sorghum computed by the profile Bowen ratio. Agron J 66:35–41

Alvo P, Desjardins RL, Schuepp PH, MacPherson JI (1984) Measurements of CO$_2$ exchange over various ecosystems. Boundary-Layer Meteorol 29:167–183

Anderson DE, Verma SB (1985) Turbulence spectra of CO$_2$, water vapor, temperature and wind velocity fluctuations over a crop surface. Boundary-Layer Meteorol 33:1–14

Austin LB, Schuepp PH, Desjardins RL (1987) The feasibility of using airborne CO$_2$ flux measurements for the imaging of the rate of biomass production. Agric For Meteorol J 39:13–23

Bach W (1978) The potential consequences of increasing CO$_2$ levels in the atmosphere. In: Williams J (ed) Carbon dioxide, climate and society. Pergomon Press, Oxford, pp 141–168

Baldocchi DD, Verma SB, Roserberg NJ (1981) Mass and energy exchanges of a soybean canopy under various environmental regimes. Agron J 73:706–710

Baldocchi DD, Verma SB, Matt DR, Anderson DE (1986) Eddy-correlation measurements of carbon dioxide efflux from the floor of a deciduous forest. J Appl Ecol 23:967–975

Baldocchi DD, Hicks BB, Meyers TP (1988) Measuring biosphere–atmosphere exchanges of biologically related gases with micrometeorological methods. Ecology 69:1331–1340

Baumgartner A (1969) Meteorological approach to the exchange of CO_2 between the atmosphere and vegetation, particularly forest stands. Photosynthetica 3:127–149

Beardsmore DJ, Pearman GI (1987) Atmospheric carbon dioxide measurements in the Australian region: data from surface observatories. Tellus 398:42–66

Benech B, Durand P, Druilhet A (1987) A case study of a non-homogeneous boundary layer (AUTAN 84 experiment). Ann Geophys 5:451–460

Bingham GE, Gillespie CH, McQuaid JH (1978) Development of a miniature, rapid response CO_2 sensor. Lawrence Livermore Natl Lab, Rep UCRL-52440

Biscoe PV, Scott RK, Monteith JL (1975) Barley and its environment III. Carbon budget of the stand. J Appl Ecol 12:269–293

Bolin B (1977) Changes of land biota and their importance for the carbon cycle. Science 196:613–615

Brach EJ, Desjardins RL, Amour G St (1981) Open-path CO_2 analyzer. J Phys E Sci Instrum 14:1415–1419

Broecker W, Ledwell JR, Takahashi T, Weiss R, Merlivat L, Memery L, Peng TH, Jahne B, Munnich KO (1986) Isotopic versus micrometeorologic ocean CO_2 fluxes: a serious conflict. J Geophys Res 91:10517–10527

Brown S, Lugo AE (1981) The role of the terrestrial biota in the global CO_2 cycle. Symp carbon dioxide issue. Div Petroleum Chemistry Inc. Am Chem Soc 26:1019–1025

Brutsaert W (1984) Evaporation into the atmosphere: theory, history and applications. D. Reidel, Lancaster, UK, 299 pp

Businger JA (1986) Evaluation of the accuracy with which dry deposition can be measured with current micrometeorological techniques. J Climate Appl Meteorol 25:1100–1123

Chahuneau F, Desjardins RL, Brach EJ, Verdon R (1989) A micrometeorological facility for eddy flux measurements of CO_2 and H_2O. J Atmos Ocean Tech 6:193–200

Christy AL, Porter CA (1982) Canopy photosynthesis and yield in soybean. In: Govindjee G (ed) Photosynthesis—applications to food and agriculture. Academic Press, New York, pp 449–511

Denmead OT, Bradley EF (1987) On scalar transport in plant canopies. Irrig Sci 8:131–149

Desjardins RL (1985) Carbon dioxide budget of maize. Agric For Meteorol J 36:29–41

Desjardins RL, Lemon ER (1974) Limitations of an eddy correlation technique for the determination of the carbon dioxide and sensible heat fluxes. Boundary-Layer Meteorol 5:475–488

Desjardins RL, MacPherson JI (1989) Aircraft-based measurements of trace gas fluxes. In: Andrea MO, Schimel DS (eds) Exchange of trace gases between terrestrial ecosystems and the atmosphere. Dahlem Konferenzen. John Wiley & Sons, Chichester, pp 138–152

Desjardins RL, Thurtell GW (1970) CO_2 exchange by an eddy correlation method. Annu Prog Rep microclimate investigations. Ithaca, NY, pp 109–112

Desjardins RL, Allen LM Jr, Lemon ER (1978) Variations of carbon dioxide, air temperature, and horizontal wind within and above a maize crop. Boundary-Layer Meteorol 14:369–380

Desjardins RL, Brach EJ, Alvo P, Schuepp PH (1982) Aircraft monitoring of surface carbon dioxide exchange. Science 216:733–735

Desjardins RL, MacPherson JI, Alvo P, Schuepp PH (1985) Measurements of turbulent heat and CO_2 exchange over forests from aircrafts. In: Hutchison BA, Hicks BB (eds) The forest atmosphere interaction. D Reidel, Dodrecht, Holland, pp 645–658

Desjardins RL, MacPherson JI, Betts AK, Schuepp PH, Grossman R (1988) Divergence of CO_2, latent heat and sensible heat fluxes: a case study. Proc lower tropospheric profiling: needs and technologies. Boulder, pp 71–72

Desjardins RL, MacPherson JI, Schuepp PH (1989a) Long wavelength contributions to aircraft-based flux estimates of CO_2, H_2O and sensible heat. In: Preprint vol 19, Conf Agriculture and forest meteorology. AMS Charleston, NC, March 7–10, pp 125–128

Desjardins RL, MacPherson JI, Schuepp PH, Karanja F (1989b) An evaluation of airborne eddy flux measurements of CO_2, water vapor and sensible heat. Boundary-Layer Meteorol 47:55–70

Dyer AJ, Hicks BB (1970) Flux-gradient relationships in the constant flux layer. QJ Meteorol Soc 96:715–721

Elagina LG, Lazarev AI (1984) Power spectrum measurements of CO_2 turbulent fluctuations in the atmospheric surface layer. Acad Sci USSR, Inst Atmos Phys (Moscow) Oceanic Physics 20:536–540 (in Russian)

Field C, Berry JA, Mooney HA (1982) A portable system for measuring carbon dioxide and water vapor exchange of leaves. Plant Cell Environ 5:179–186

Francey RJ, Garratt JR (1981) Interpretation of flux-profile observations at ITCE, 1976. J Appl Meteorol 20:603–618

Fuchs M, Tanner CB (1970) Error analyses of Bowen ratio measured by differential psychrometry. Agric Meteorol 7:329–334

Goudriaan J (1977) Crop micrometeorology: a simulation study. Cent Agric Publ Doc (Wageningen) Annu Rep, The Netherlands, 249 pp

Greenhut GK, Bingham GE, Gilmer R (1983) Aircraft measurements of the flux of CO_2 over the Caribbean Sea. In: Symposium 19. The ocean and the CO_2 climate response. IUGG Meet, Hamburg, pp 15–27

Grossman RL, Bean BR (1973) An aircraft investigation of turbulence in the lower layers of a marine boundary layer. NOAA Tech Rep ERL 291-WMPO, No 4, 166 pp

Heikinheimo MJ (1986) Techniques for detecting carbon dioxide and water vapor transport above a vegetated surface using the eddy correlation method. PhD Thesis, Guelph Univ, Canada, 206 pp

Hicks BB (1986) Measuring dry deposition: a re-assessment of the state of the art. Water Air Soil Pollut 30:75–90

Houghton RA, Woodwell GM (1980) The flax pond ecosystem study: exchanges between a salt marsh and the atmosphere. Ecology 6:1434–1445

Huber B (1952) Der Einfluss der Vegetation auf die Schwankungen des CO_2 Gehaltes der Atmosphäre. Arch Meteorol Geophys Blioklimatol Ser B 4:154–167

Inoue E, Tani EN, Imai K, Isobe S (1958) The aerodynamic measurement of photosynthesis over a nursery of rice plants. J Agric Meteorol Tokyo 14:45–53

Inoue E, Vchizima Z, Sarto T, Isobe S, Vemura K (1969) The "assimitron", a newly devised instrument for measuring CO_2 flux in the surface air layer. J Agric Meteorol 25:165–170

Jones EP, Smith SD (1977) A first measurement of sea-air CO_2 flux by eddy correlation. J Geophys Res 82:5990–5992

Jones EP, Zwick H, Ward TV (1978) A fast response atmospheric CO_2 for eddy correlation flux measurement. Atoms Environ 12:845–851

Kanemasu ET, Powers WL, Sij JW (1974) Field chamber measurements CO_2 flux from soil surface. Soil Sci 118:233–237

Kanemasu EG, Wesely ML, Hicks BB, Hettman JL (1979) Techniques for calculating energy and mass fluxes. In: Barfield BJ, Gerber JF (eds) Modification of the aerial environment of plants. American Society of Agricultural Engineers. St. Joseph, MI, pp 156–182

Keller M, Kaplan WA, Wofsy SC (1986) Emissions of N_2O, CH_4, CO_2 from tropical forests soils. J Geophys Res 91:11791–11802

Lemon ER (1960) Photosynthesis under field conditions. II. An aerodynamic method for determining the turbulent carbon dioxide exchange between the atmosphere and a corn field. Agron J 52:697–703

Lemon ER (1967) Aerodynamic studies of CO_2 exchange between the atmosphere and the plant. In: San Pietro A, Greer FA, Army TJ (eds) Harvesting the sun: photosynthesis in plant life. Academic Press, New York, pp 263–290

Lemon ER (1969) Important microclimatic factors in soil–water–plant relationships. Modifying the soil and water environment for approaching the agricultural potential of the Great Plains. Great Plains Agric Council Publ 3:95–102

Lemon ER, Wright JL (1969) Photosynthesis under field conditions. XA. Assessing sources and sinks of carbon dioxide in a corn (Zea mays L.) crop using a momentum balance approach. Agron J 61:405–411

Lenschow DH, Stankov BB (1986) Length scales in the convective boundary layer. J Atmos Sci 43:1198–1209

Lenschow DH, Delany AC, Stankov BB, Stedman DH (1980) Airborne measurements of the vertical flux of ozone in the boundary layer. Boundary-Layer Meteorol 19:249–265

Leuning R, Denmead OT, Lang ARG, Ohtaki E (1982) Effects of heat and water vaport transport on eddy covariance measurement of CO_2 fluxes. Boundary-Layer Meteorol 23:209–222

Mack AR, Desjardins RL, MacPherson JI, Schuepp PH (1989) Relative photosynthetic activity of agricultural lands from airborne carbon dioxide and satellite data. Int J Remote Sens 11:237–252

MacPherson JI, Morgan JM, Lum K (1981) The NAE Twin Otter atmospheric research aircraft. Natl Res Counc Can Rep LTR-FR-80. Ottawa, Canada, 21 pp

MacPherson JI, Desjardins RL, Schuepp PH (1987) Gaseous exchange measurements using aircraft-mounted sensors. 6th Symp meteorological observations and instrumentation. New Orleans Jan 12–16, pp 128–131

McBean GA, Paterson RD (1975) Variations of the turbulent fluxes of momentum, heat and moisture over Lake Ontario. J Phys Oceanogr 5:523–531

Meinguet J (1979) Multivariate interpolation at arbitrary points made simple. J Appl Math Phys 30:292–304

Monteith JL, Szeicz G (1960) The carbon dioxide flux over a field of sugar beets. Q J R Meteorol Soc 86:205–214

Mooney HA, Vitousek PM, Matson PA (1987) Exchange of materials between terrestrial ecosystems and the atmosphere. Science 238:926–932

Norman JM (1979) Modeling the complete canopy. In: Barfield B, Gerber J (eds) Modification of the aerial environment of crops. Am Soc Agric Engr, Monogr No 2, ASAE, St. Joseph, MI, pp 249–277

Norman JM, Hesketh JD (1980) Micrometeorological methods for predicting environmental effects on photosynthesis. In: Hesketh JD, Jones JW (eds) Predicting photosynthesis for ecosystem models, vol 1. CRC Press, Boca Raton, pp 9–35

Ohtaki E (1984) Application of an infrared carbon dioxide and humidity instrument to studies of turbulent transport. Boundary-Layer Meteorol 29:85–107

Ohtaki E (1985) On the similarity in atmospheric fluctuations of carbon dioxide, water vapor and temperature over vegetated fields. Boundary-Layer Meteorol 32:25–37

Ohtaki E, Matsui T (1982) Infrared device for simultaneous measurements of fluctuations of atmospheric carbon dioxide and water vapor. Boundary-Layer Meteorol 24:109–119

Ohtaki E, Seo T (1976) Infrared device for measurement of carbon dioxide fluctuations under field conditions. II. Double beam system. Berichte des Ohara Instituts für Landwirtschaftliche Biologie, vol 16. Okayama Univ, pp 183–190

Pearson CJ, Larson EM, Hesketh JD, Peters DB (1986) Development and source-sink effects of single leaf and canopy carbon dioxide exchange in maize. Field Crops Res 9:391–402

Peng TH, Takahashi T, Broecker WS (1974) Surface radon measurements in the North Pacific Ocean Station Papa. J Geophys Res 79:1772–1780

Peterson RB, Zelitch I (1982) Relationship between net CO_2 assimilation and dry weight accumulation in field grown tobacco. Plant Physiol 70:677–685

Pruitt WO, Morgan DL, Lourence FJ (1973) Momentum and mass transfer in the surface boundary layer. Q J R Meteorol Soc 99:370–386

Roether W (1980) The effect of the ocean in the global carbon cycle. Experientia 36:1017–1134

Saïd F (1988) Etude experimentale de la couche limite marine: structure turbulente et flux de surface (Expérience TOSCANE-T). PhD Thesis, Université Paul Sabatien, Toulouse, 316 pp

Saugier B (1976) Sunflower. In: Monteith JL (ed) Vegetation and the atmosphere. Academic Press, London, pp 87–120

Schuepp PH, Desjardins RL, MacPherson JI, Boisvert JB, Austin LB (1987) Airborne determination of regional water use efficiency and evapotranspiration: present capabilities and initial field tests. Agric For Meteorol 41:1–19

Schuepp PH, Desjardins RL, MacPherson JI, Boisvert JB, Austin LB (1989) Interpretation of airborne estimates of evapotranspiration. In: Estimations of Areal Evapotranspiration (Proc Worksh Vancouver, BC, Aug 1987. IAHS Publ No 177, pp 185–196

Schuepp PH, Leclerc MY, MacPherson JI, Desjardins RL (1989b) Footprint prediction of scalar fluxes from analytical solutions of the diffusion equation. Boundary-Layer Meteorol 50:355–373

Sellers PJ, Hall FG, Asrar G, Strebel DE, Murphy RE (1988) This First ISLSCP field experiment (FIFE). Bull Am Meteorol Soc 69:22–27

Shawcroft RW, Lemon ER, Allen LH, Stewart DW, Jensen SE (1974) The soil–plant–atmosphere model and some of its applications. Agric Meteorol 14:287–307

Sinclair TR, Allen LH Jr, Lemon ER (1975) An analysis of errors in the calculation of energy flux densities above vegetation by Bowen ratio-profile method. Boundary-Layer Meteorol 8:129–139

Sinclair TR, Murphy CE Jr, Knoerr KR (1976) Development and evaluation of simplified models for simulation canopy photosynthesis and transpiration. J Appl Ecol 13:813

Smith SD, Jones EP (1986) Isotopic and micrometeorological ocean CO₂ fluxes: different time and space scales. J Geophys Res 91:10529–10532

Smithie WM, Takahashi T, Chipman DW, Ledwell JR (1985) Gas exchange and CO₂ flux in the tropical Atlantic Ocean determined from ²²²Rn and; pCO₂ measurements. J Geophys Res 90:7005–7022

Stewart DW, Dwyer LM (1986) Development of a growth model for maize. Can J Plant Sci 66:267–280

Tanner CB, Thurtell GW (1969) Anemoclinometer measurements of Reynolds stress and heat transfer in the atmospheric surface layer. Final Rep TRECOM 66-G22-F, Univ Wisconsin, Madison, pp R1–R10

Tucker CJ (1979) Red and photography infrared linear combination for montitoring vegetation. Remote Sens Environ 8:127–150

Verma SB, Rosenberg NJ (1975) Accuracy of lysimetric, energy balance and stability corrected aerodynamic methods of estimating above-canopy flux of CO₂. Agron J 67:699–704

Verma SB, Baldocchi DD, Anderson DE, Matt DR, Clement RJ (1986) Eddy fluxes of CO₂, water vapor, and sensible heat over a deciduous forest. Boundary-Layer Meteorol 36:71–91

Verma SB, Kim J, Clement RJ (1989) Carbon dioxide, water vapor and sensible heat fluxes over a tall grass prairie. Boundary-Layer Meteorol 46:53–67

Volkov Yu A, Elagina LG, Lazarev AI, Lomadze SO (1986) Simultaneous measurements of moisture and carbon dioxide fluxes in the atmosphere near the earth. Izv, Atmos Oceanic Phys 22:591–595

Webb EK, Pearman GI, Leuning R (1980) Correlation of flux measurements for density effects due to heat and water vapour transfer. Q J R Meteorol Soc 106:85–100

Wesely ML (1988) Use of variance techniques to measure dry air surface exchange rates. Boundary-Layer Meteorol 44:13–31

Wesely ML, Cook DR, Hart RL, Williams RM (1982) Air–sea exchange of CO₂ and evidence for enhanced upward fluxes. J Geophys Res 87:8827–8832

Wyngaard JC (1986) Measurement physics. In: Lenschow DH (ed) Probing the Atmospheric Boundary Layer. AMS, Boston, pp 5–18

Wyngaard JC (1988) Flow-distortion effects on scalar flux measurements in the surface layer: Implications for sensor design. Boundary-Layer Meteorol 42:19–26

Wyngaard JC, Zhang S (1985) Transducer-shadow effects on turbulence spectra measured by sonic anemometer. J Atmos Ocean Tech 2:548–558

Interaction of Carbon Dioxide with Growth-Limiting Environmental Factors in Vegetation Productivity: Implications for the Global Carbon Cycle

R.M. Gifford

Contents

Symbols and Acronyms

ppm	For carbon dioxide, parts per million by volume in the atmosphere
NPP	Net primary production of vegetation ($t\,ha^{-1}\,yr^{-1}$)
R_n	Net radiation
λ	Latent heat of evaporation of water
P	Annual precipitation
P_n	Net photosynthetic CO_2 fixation per unit area of leaf
Rubisco	Ribulose bisphosphate carboxylase/oxygenase
V_x	Rate of Rubisco oxygenase reaction at saturating O_2 concentration
V_c	Rate of Rubisco carboxylase reaction at saturating CO_2 concentration
K_x	Michaelis constant for Rubisco oxygenation
K_c	Michaelis constant for Rubisco carboxylation
C_i	CO_2 concentration within the leaf mesophyll
X	Oxygen concentration within the leaf mesophyll
J	Rate of absorption of photosynthetically active photons by photosynthetic tissues

1 Introduction

Atmospheric carbon dioxide concentration has increased globally from about 280 ppm before the Industrial Revolution (Pearman 1988) to about 353 ppm in 1990. That increase, and the continuing increase at a rate of about 1.5 ppm per annum, owing mainly to fossil fuel burning, is likely to cause change in climate, in primary productivity of terrestrial vegetation (managed and unmanaged), and in the degree of net sequestration of atmospheric CO_2 into organic form. The quantitative role of the latter in attenuating the increase in atmospheric CO_2 concentration itself is an important but uncertain element of the global carbon-cycle models that are required to predict future increases of atmospheric CO_2 concentration.

In my experience in workshops and other multidisciplinary gatherings, argument arises in discussion of this topic among different groups of scientists such as bioclimatologists, plant physiologists, biogeochemists and ecologists. Plant physiologists are often impressed by the positive effect of higher CO_2 concentration on plant growth under experimental controlled environments and argue that this would be at least partly expressed in the field for many species and communities. Others being sensitive to the limitations to productivity imposed by other environmental factors, assert that these preclude long-term CO_2 responsiveness of vegetation growth. Reference is also often made to the progressive attenuation of effects observed at low levels of biological organization, such as the leaf or isolated plant levels, when moved to successively higher levels such as the field over a full year (Gifford 1974a), i.e. to the question of scaling up from low-level observations to higher-level real-world conditions. Bioclimatologists interested in global atmospheric change tend to focus on prospective change in temperature and rainfall, ignoring the certain prior change in CO_2 concentration and its direct effects on plants. Some ecologists emphasize tropical deforestation, suggesting that global biomass must be getting smaller.

In this review, emphasis is placed on the interactions of CO_2 concentration and vegetation properties in the global carbon cycle.

2 Scales of, and Limitations to, Vegetation Productivity

There are many elements to disagreements on scales and limitations, but there are three aspects which recur.

1. Definition of "Limitation" and the Distinction
Between Endogenous Regulation and External Inputs

In some ecological discussions it has been implicitly assumed that limitation is "all or nothing", i.e. that a plant that is limited by an external factor will respond by increasing its growth by 1% for each 1% increase in availability of that external factor. Part of this idea is the assumption that if growth is limited by one factor it cannot be limited by another at the same time. It seems to derive from an extreme interpretation of Liebig's (1855) "law of the minimum". This conceptual or intuitive preference is also strong among some biochemists in relation to endogenous metabolic control of biochemical pathways: e.g. following some discussion of this matter, in the context of regulation of photosynthesis and C-partitioning within the leaf, in recent years among plant biochemists, it was observed in one paper (Stitt and Quick 1989) that:

"The question naturally follows, which of these [enzymatic steps] makes the larger contribution to controlling the rate of sucrose synthesis. Obviously, at steady state, every enzyme in a pathway will be catalyzing the same net flux but, intuitively, *it is still natural to feel that only one (or a few) are involved in determining the magnitude of the flux*, i.e. are in control" (my italics).

Perhaps there is evolutionary advantage of parsimony in the design of endogenous regulation of metabolic pathways [although the concept of "bottlenecks" or "pacemakers" in metabolic regulation has been challenged (Kacser and Burns 1973)]. However, it is not obvious why vegetation would deploy itself, in its evolved approach to exogenous resource acquisition, such that its growth is responsive to increase in supply of only one resource at a time except under extreme conditions. For, if a single resource were alone limiting, there would be selection pressure leading to genetically modified metabolism, development and/or habit for acquiring less of the abundant resources and in favour of harnessing the limiting resource.

On the assumption that more than one endogenous process or external factor at a time can influence performance, a more realistic model is to define a degree of limitation as follows (Gifford 1974b):

"The rate of process A is limited by the rate of process B (or level of factor C) if an increment in the rate of process B (or level of factor C) leads to some increment in the rate of A".

This definition has similarity to concepts that were emerging at that time in metabolic regulation under the title of "control analysis" (Kacser and Burns 1973). This is a technique intended to partition internal factors controlling rates of metabolic processes into additive components having "sensitivity coefficients" that sum to unity thereby accounting for the full internal limitation to a process given the prevailing environment.

It is appropriate to distinguish clearly whether one is dealing with external inputs to a system or with endogenous controls to the system. In this review we are concerned with CO_2 as an external limiting factor. I am not aware of any extension to control theory dealing with external factors that influence plant productivity. Many ecosystem models deal with external limiting factors in one of two ways. Either they use the "law of the minimum", in which the supposed most limiting external factor paces a process, or they use a multiplicative model in which some measure of degrees of limitation by several colimiting external factors are multiplied to give a composite limitation. Neither reflect reality very well. The appropriateness of each, as an approximation of convenience, depends on the time scale over which the system is being modelled, on the scope for varying external factors, on the relative magnitude of the colimitations, and on the nature of the interactions between them.

2. Colimiting Factors Can Interact in Their Effects

Complex interaction in the effects of colimiting factors means that it may not be meaningful to attempt to define limitations such that the sum of partial limitations by all colimiting external factors must add up to unity, or to assume that the product of colimitations is an appropriate approach. Owing to positive feedbacks and synergisms between factors, alleviation of two limitations concurrently might increase growth by more than the sum of their individual effects. Physiological adaptations in plants occurring over time can compensate for restrictive levels of some exogenous inputs (Bloom et al. 1985) such that the impact of several limitations to growth become similar (Chapin et al. 1987).

Consider a hypothetical plant community that is limited concurrently by two external factors, rainfall and a certain soil nutrient that is becoming exhausted in the rooting zone but still present at depth. Suppose also that there is no permanent water table within root range (i.e. the vegetation is reliant solely on seasonal rainfall), and that the volume of the rooting zone is dictated by the maximum depth that the soil water penetrates. If an increasing annual rainfall increased the wetted depth and therefore the availability of previously untapped soil for root exploration, then the effect of increased rainfall would ultimately be greater than the immediate direct impact, as roots would grow deeper and gradually mobilize new reserves of the scarce nutrient. This would be a positive feedback, developing over time. A similar synergism could occur between water and nitrogen, where the nitrogen taken up is predominantly that which is mineralized to soluble forms (NO_3^- and NH_4^+) from unavailable soil organic N. Dryland cereal production is usually water-limited, i.e. productivity correlates strongly with precipitation. However, organic matter mineralization and hence available nitrogen is also closely related to soil moisture. So the yield response to water application may in many instances be a response to water and nitrogen acting together. However, with continued irrigation over years the soil organic nitrogen level would reach a new equilibrium and that source of yield stimulation would become exhausted. These examples introduce the third probable source of disagreements about limiting factors in understanding the impacts of global atmospheric change, namely the time frame of consideration.

3. Time Frame over Which Limitations May Be Manifested

Most controlled plant experiments are performed on plants or plant parts on the scales of centimeters and minutes. Fewer are done on the scale of meters and days (e.g. whole plant diurnal carbon balance), or square meters to hectares and weeks to years (community growth). Carbon dioxide enrichment experiments cannot be conducted on the biggest scales (continental to global) for which we need answers for purposes of socio-economic policy relating to global atmospheric composition change. We must therefore develop models of the carbon cycle for policy analysis and decisions, based on observations and simplifying abstractions on large scales linked with experiments on smaller scales.

3 The Global Carbon Cycle

The global carbon cycle is in fact many nested cycles with turnover times ranging from seconds (photosynthesis–photorespiration cycle) to hundreds of millions of years (the carbonate–silicate cycle). Fast turnover cycles can involve small pools but large global fluxes and turnover per year. The slow cycles involve large pools but small global fluxes. The fast cycle, involving CO_2 fixation by biospheric gross photosynthesis and decarboxylation by respiration and photorespiration, probably turns over more than 120 Gt(carbon) year^{-1} of the small atmospheric pool of about 740 Gt(C) (Gifford 1982). At the other extreme, the slow geochemical carbonate–silicate cycle turns over only about 0.1 Gt(C) year^{-1} (or less) of the vast lithospheric pool of about 65×10^6 Gt(C) (Bolin 1983).

There are interactions between the carbon subcycles. Vegetation productivity is involved across the full range, probably even in the grand carbonate–silicate cycle that is traditionally (Berner et al. 1983) viewed as a purely geochemical process setting outer bounds between which atmospheric CO_2 concentration can fluctuate in the very long term.

Consideration of probable biological involvement in the latter geological mega-cycle is important for understanding the faster aspects of the carbon cycle because, in setting the coarse limits of atmospheric CO_2 concentration, this cycle also strongly influences the Earth's average temperature and hence the broad limits of the planet's climate within which the biological systems (and fast-turnover subcycles) operate. The global carbon cycle involves carbon turnover and partitioning between the atmosphere, the biosphere, the hydrosphere, the lithosphere and magma. More than 99.9% of the C is in the Earth's crust and magma, less than 0.1% in the hydrosphere, atmosphere, soil and biosphere combined (Bolin 1983). The partitioning is determined by the balance between rate of terrestrial rock weathering and the rate of tectonic turnover of the crust (Broecker 1973; Walker et al. 1981). It is thought to work as follows.

Weathering by carbonic acid in rain water and soil water releases Ca^{2+} and Mg^{2+} ions from the rocks and soil. These solubilized ions flow (as does dissolved silicon dioxide) in the rivers with bicarbonate to the ocean where marine organisms turn the bicarbonate ions into calcium and magnesium carbonate shells. Some of this carbonate accumulates (along with a much lesser amount of organic matter) as ocean floor sediments that become sedimentary rock which is ultimately subducted into the magma (along with silicon dioxide) at descending tectonic plate boundaries. Carbon is thereby removed from the atmosphere via the ocean buffer to the crust and magma. It is ultimately returned to the atmosphere/ocean pools as CO_2 via volcanic outgassing, the $Ca(Mg)CO_3$ and SiO_2 having been converted under high temperature and pressure back into calcium (magnesium) silicate rock by magmatic metamorphosis (Berner et al. 1983). If the rate of rock weathering over this geologic time scale (10^8 years) changes, then the dynamic atmospheric balance between removal by carbonate sedimentation and release via magmatic outgassing, and hence atmospheric CO_2 content, gradually changes.

Calcium carbonate precipitation in the ocean is photosynthesis-dependent for two reasons. Much, or most, carbonate is made as protective shells of organisms (Holland et al. 1986) and therefore depends on the level of ocean photosynthetic productivity. Secondly, and more importantly, the rate of rock weathering is dependent on CO_2 concentration in the soil, which is determined by root respiration and microbial decomposition of soil organic matter, as well as by organic acid exudation from roots. The soil has a 10- to 100-fold higher CO_2 concentration than is present in the atmosphere (Holland et al. 1986). The rate of root respiration and soil organic matter decomposition is related to the rate of shoot and root litter deposited by plants which for mature ecosystems must approximately equal net primary productivity. If net primary production of the terrestrial biosphere were CO_2 dependent, then this would complete a negative feedback loop in the grand carbonate–silicate cycle (Volk 1987). This would ultimately partition into rocks and magma the atmospheric CO_2 "bleep" caused by man – probably long after our civilization or even after *Homo sapiens* has disappeared.

Thus, biological carbon turnover is as critically involved as is physical lithosphere turnover and "chemical" weathering in determining the quantitative scale of the global carbonate–silicate cycle and hence atmospheric composition and climate (via the greenhouse effect) on geologic time scales. For example, Volk (1989) has presented evidence that the substantial displacement, over the last $10^7 - 10^8$ years, of gymnosperms by angiosperms (which cause about three times faster rock-weathering rates than gymnosperms owing to the deciduous lifeform) has caused a substantial reduction in atmospheric CO_2 concentration via the carbonate–silicate cycle. He hypothesized that this has contributed to the overall climatic cooling during that period culminating in the present series of ice ages over the last 2×10^6 years.

Over the full 3 to 4×10^9 years that life has been evolving on Earth, it has been responsible for a net deposit of over 60×10^6 Gt of carbon into sedimentary carbonate and organic matter in rock. That is, the biosphere has "pumped" more than 80 000 times as much carbon out of the atmosphere into sedimentary rocks as is currently airborne. This factor accounts just for the net accumulation and does not include the subsequently recycled C from magmatic CO_2 emissions derived from subducted sediments. The original source of most of this CO_2 presumably was magmatic outgassing which must have occurred at an average net rate of about 0.02 Gt(C) per year over the life of the Earth to account for the net sedimentary accumulation of carbon. Original CO_2 partial pressure in the atmosphere was certainly many times higher than it is today, possibly exceeding the present total atmospheric pressure of 1 bar (Schwartzman and Volk 1989). Thus, photosynthesis (as the driver behind the source of soil carbon dioxide and acid root exudates that accelerate rock weathering) in concert with biological shell formation has restricted atmospheric CO_2 to a low level despite a huge cumulative amount of CO_2 seeping from the Earth. The rate of weathering and calcium carbonate precipitation would have been much less, perhaps 10 to 1000 times less (Lovelock and Whitfield 1982; Schwartzman and Volk 1989) in the absence of the life. Hence, atmospheric CO_2 concentration would be much higher than it now is. Surface temperature would probably be 15–45 °C higher than it is today on an abiotic Earth (Schwartzman and Volk 1989).

Since the biosphere, acting as an amplifier of the carbonate–silicate cycle, has kept atmospheric CO_2 down to a level of about 180–340 ppm for at least the last 160 000 years (Barnola et al. 1987), it seems very likely that it will continue to do so on a geologic time scale despite anthropogenic CO_2 emissions from the relatively minute pool of fossil fuels. However, since the rate of fossil fuel burning [5.4 Gt(C) yr^{-1}] is now about 50 times faster than the rate at which the calcium–silicate cycle removes it, it will take a long time for the carbonate–silicate cycle to catch up after fossil fuel combustion ceases after several centuries.

However, the question remains whether, on a shorter time scale of significance to man's energy and land management policy decisions (i.e. decades), the biosphere is contributing to net carbon removal from the atmosphere. In the short term, the mechanism of carbon removal by terrestrial vegetation would not be by stimulation of weathering, but rather by an increase of organic matter storage in intermediate pools of standing biomass and soil organic matter. In fact, an increase in the soil organic pool may well be a prerequisite to the enhanced microbial activity leading to the increased weathering rates that underpin the very much slower-moving carbonate–silicate mechanism of carbon sequestration. However, it is not necessarily just the CO_2 fertilizing effect on photosynthesis that could cause such increased organic matter deposition. Any associated increase in temperature and rainfall (the greenhouse effect) would also be expected to increase biosphere productivity and carbon storage. For the paleoclimatic record indicates that during warm periods in the past, as seen in glacial-interglacial comparisons, openlands and grasslands have given way to encroaching forest which receded in cool periods (Singh 1988). Further, examination of the current productivity of the world's ecosystems indicates that annual productivity shows broad positive correlations with both temperature and rainfall (Leith 1972). However, in terms of the time scale of a century, the approximately 0.5 °C global average temperature increase has not yet unquestionably been shown to have exceeded the range of "normal" variability, while CO_2 concentration has increased 25% from 280 to 353 ppm (Kuo et al. 1990). Thus, the CO_2 factor requires prior emphasis over temperature and rainfall change in seeking to predict changes to the global carbon cycle on the decadal time scale.

4 Vegetation Productivity in Relation to Atmospheric CO_2

Detailing the mechanisms of CO_2 effects on productivity and soil organic matter in the face of other limiting factors is a complex question for which data are inadequate. Since adequately replicated, direct long-term controlled CO_2 experimentation will be only rarely, if ever, achieved in the field, even with "free air CO_2 experiment" equipment (Hendrey et al. 1988), other approaches are required to evaluate quantitatively the likely impact of the globally increasing CO_2 concentration on primary production and organic matter turnover. Mechanistic modelling based on experimental investigation of simplified systems in controlled environments is the most obvious way ahead.

A critical feature of such models must be the correct representation of the interaction between CO_2 concentration and the other environmental variables that colimit vegetation productivity. As illustrated in Sect. 2, concepts based on the "law of the minimum" or on multiplication of colimitations can fall far short of reality.

To model the incremental sequestration of carbon into live and dead organic matter, the impact of CO_2 on annual productivity is only a start. Annual productivity is the input to the interlocking cycles of carbon and mineral nutrient turnover in the plant/soil system. Turnover involves organic matter reoxidation by (1) herbivory by both invertebrates and vertebrates; (2) decomposition by decomposer organisms;

and (3) fire. All three may be influenced by both atmospheric CO_2 concentration and any future climate change. For example, high CO_2 concentration might increase the carbon to nitrogen ratio as well as the size of plants. Insects must process a lot of unneeded carbon to acquire enough nitrogen. A higher C:N ratio would increase the amount of tissue an insect must eat to acquire its N with consequential adverse effects on fecundity. It is commonly believed that insect population growth is inversely related to plant vigor (Mattson and Addy 1975) and directly related to %N content (White 1984). Thus, one can reasonably hypothesize that plants grown in high CO_2 concentrations will experience less herbivory owing to lower populations of herbivores and therefore more litterfall of higher C:N ratio material. This would represent an enhanced carbon sequestration into the litter pool. Such a phenomenon could in turn lead to two further repercussions. The enhanced buildup of plant litter perhaps could lead to more frequent or more intense fires. These would have a complex impact on C-storage. Another possibility is that a buildup of high C:N ratio litter, and hence soil organic matter, could provide the carbon substrate to foster enhanced non-symbiotic nitrogen fixation. Alternatively, it might lead to the conversion of more phosphate to organic phosphates which are unavailable for plant growth. Such considerations are beyond the scope of this review for detailed evaluation. They lie further down the track than the question which is addressed here, namely: "Does the increasing atmospheric CO_2 concentration increase the primary productivity of field vegetation that is colimited by the levels of several other environmental factors?"

5 Interaction of CO_2 and Other Growth-Limiting Environmental Variables

Annual productivity of vegetation is usually constrained by one or more of the following environmental variables: (1) incident radiation; (2) water supply; (3) temperature; (4) one or more mineral nutrients such as nitrogen or phosphate; (5) adverse soil conditions such as extreme pH, salinity, waterlogging or compaction. Colimitation by more than one factor is likely to be the norm because of the plant's morphological and physiological flexibility in acquiring resources. That is, plants are likely to adjust their resource acquisition strategy until several inputs are about equally colimiting when limitations are ascertained in the short term (Chapin et al. 1987). There are limits to this flexibility, however. For example at constant CO_2 concentration, where water and nutrients are abundant and temperature is optimal, variation in incident radiation is likely to be a pacesetter of variation in productivity. Although the vegetation can develop during the growing season in such a way as to increase its light interception, where the leaf canopy intercepts all available light, there is little it can adjust to capture more light per unit land area. Similarly, where temperature is very low, thereby reducing the rate of chemical interactions and hence sink growth, there is relatively little that the plant can do to increase the heat input by investing in greater use of less restrictive resources. So low temperature would remain a dominant pacesetter of growth. Is it possible, however, that a higher CO_2 concentration would allow the plant to compensate for the impact of growth-limiting levels of other variables?

5.1 The Principal Causes of Variation in Vegetation Productivity

Annual incident radiation is often a limitation to plant growth. Most productivity models use some criterion of the radiation regime as the main driver of productivity, with some criterion of rainfall being the principal modifier of the productivity vs radiation relationship. Temperature is often modelled as determining the rate of phenological development.

For example, using International Biological Program (IBP) data gathered from the world's climatic zones, Uchijima and Seino (1985, 1987) developed the semi-empirical Chikugo model which relates annual net primary productivity (NPP) to net radiation, R_n, with the proportionality coefficient being a function of the ratio between annual net radiation and precipitation:

$$NPP = \{0.29[\exp(-0.216(Rn/\lambda P)2)]\}R_n, \tag{1}$$

where P is annual precipitation and λ is the latent heat of evaporation.

This relationship is shown in Fig. 1, with the combinations of NPP, R_n and $R_n/\lambda P$ for major biomes mapped on to it (Uchijima and Seino 1987). The adequacy of such a simple model to describe biome productivity at that coarse level of resolution is indicative that the major variations in vegetation productivity around the world are attributable to variations in solar energy input and water. The absence of CO_2 in the model does not necessarily mean that productivity is not CO_2-sensitive. It merely reflects the fact that CO_2 concentration is not very spacially variable over the Earth's surface. Both the light and temperature inputs are positively related to net radiation. So to understand the role that CO_2 increase plays we must first ask how CO_2 concentration interacts with light, water and temperature and other environmental factors in determining productivity.

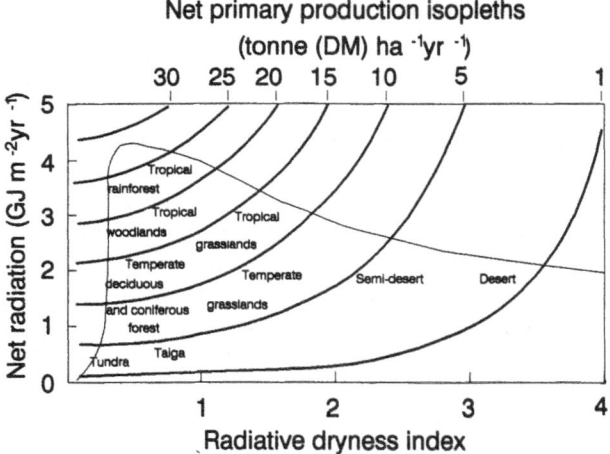

Fig. 1. Net annual primary productivity isopleths for natural ecosystems as a function of annual net radiation (R_n) and the radiative dryness index of the site according to the Chikugo model. The radiative dryness index is $Rn/\lambda P$, where P is annual precipitation and λ is latent heat of evaporation. The approximate domains occupied by the world's major biomes are superimposed (After Uchijima and Seino 1987)

5.1.1 Interaction of CO_2 with Photosynthetic Irradiance

Much vegetation grows at growth-restricting levels of photosynthetic light input. The "law of limiting factor" concept that such vegetation cannot respond to increased CO_2 concentration (Lemon 1977, Kramer 1981) does not apply to C_3 species because of the character of the interactions between photosynthesis, photorespiration and dark respiration. This can be explained by a model of photosynthesis based on knowledge of the biochemistry of photosynthesis and photorespiration (von Caemmerer and Farquhar 1981).

The primary CO_2-fixing enzyme (ribulose bisphosphate carboxylase-oxygenase, "Rubisco") catalyzes the reaction of a CO_2 molecule with ribulose bisphosphate to produce 3-phosphoglyceric acid, which is partially transformed into sucrose while ribulose bisphosphate is regenerated. In the presence of oxygen, Rubisco also catalyzes the reaction of oxygen with ribulose bisphosphate to produce phosphoglycolate which enters the photorespiratory cycle that gives off half the carbon cycling through it as CO_2, while regenerating ribulose bisphosphate from the other half. Oxygen and CO_2 are competitive inhibitors of each other's reaction with ribulose bisphosphate on the Rubisco enzyme. The light reactions of photosynthesis produce the chemical energy that drives these two linked metabolic cycles, each regenerating ribulose bisphosphate. So light limitation acts via a limited input of ribulose bisphosphate to both the carboxylase and oxygenase reactions catalyzed by Rubisco.

Net photosynthetic CO_2 fixation, P_n, is a balance between the rate of gross photosynthetic carboxylation and the rate of photorespiratory decarboxylation. From knowledge of Rubisco enzyme kinetics, the light-limited (i.e. ribulose bisphosphate-limited) net photosynthesis rate is given by (Sharkey 1986):

$$P_n = \frac{[1 - 0.5(V_x \cdot K_c \cdot X/V_c \cdot K_x \cdot C_i)]J}{[9.24 + 10.38(V_x \cdot K_c \cdot X/V_c \cdot K_x \cdot C_i)]}, \qquad (2)$$

where V_x = maximum rate of oxygenation at saturating oxygen;
V_c = maximum rate of carboxylation at saturating CO_2;
K_x = Michaelis constant for oxygenation;
K_c = Michaelis constant for carboxylation;
C_i = CO_2 concentration in solution in the mesophyll cells;
X = oxygen concentration in solution in the mesophyll cells;
J = rate of absorption of photosynthetically active photons.

In applying this equation, I assume that the ratio of CO_2 concentration inside the leaf to the atmospheric concentration, Ca, is

$$C_i/C_a = 0.7 \qquad (3)$$

at all C_a.

Equations (2) and (3) predict that increasing C_a from 350 to 700 ppm would increase light-limited net photosynthesis at 20 °C by 21%.

At a light flux that saturates net photosynthesis, ribulose bisphosphate supply in the leaf does not limit the rate, but the rate of its carboxylation may. This depends on Rubisco activity and CO_2 concentration according to the kinetic equation

(Jordan and Ogren 1984; Sharkey 1986):

$$P_n = \frac{[1 - 0.5(V_x \cdot K_c \cdot X/V_c \cdot K_x \cdot C_i)V_c \cdot C_i]}{C_i + K_c(1 + X/K_c)} \tag{4}$$

Evaluation of this equation along similar lines to that for Eqs. (2) and (3) shows a 45% theoretical relative stimulation of leaf net photosynthesis at two times normal CO_2 under light-saturating conditions at 20 °C.

The response of growth to doubled CO_2 involves not only net photosynthesis of leaves but also whole-plant respiration, R, comprising growth (g) and maintenance (m) elements:

$$R = R_m + R_g. \tag{5}$$

Growth rate (G) is equal to daytime net leaf photosynthesis, P_n, less 24-h respiration, R, i.e.

$$G = P_n - R. \tag{6}$$

It is not simple to extend the above equations to predict theoretically the response of G to CO_2 under light-limiting and light-saturating conditions. This is because one needs to be able to specify the amount of standing live biomass in relation to net leaf photosynthesis in order to compute the maintenance respiration rate component. Does the ratio of standing live biomass to net photosynthesis per unit ground area increase or decrease when plants are grown under high photon irradiance or at high CO_2? This will depend on the factors that cause tissues to die leading, at community maturity, to a more or less constant live biomass. If at low radiation levels there is a tendency for the ratio of biomass (and hence maintenance respiration) to net photosynthesis to be greater than under high radiation, then the comparative sensitivity of growth to CO_2 enrichment at low vs high radiation would be boosted at low radiation.

Another complication is that there is some evidence that, for unknown reasons, plants growing in high CO_2 concentration may exhibit up to a 10–20% reduction in the maintenance respiration per unit plant dry matter (Silsbury and Stevens 1984; Reuveni and Gale 1985; Gifford et al. 1985). This is despite such plants containing larger concentrations of soluble carbohydrates which often seem to lead to elevated rates of respiration. Assuming, for illustrative purposes, that high CO_2 concentration reduces the maintenance coefficient by 10%, that maintenance respiration is 20% of net photosynthesis at normal CO_2 concentration, and that standing live biomass remains unchanged when CO_2 concentration is doubled, then the equations and assumptions indicate that net annual growth (and hence root plus shoot litterfall) would increase in response to double CO_2 concentration by 27% when light-limited and by 50% when light-saturated.

Although the above calculations probably reflect much of the essence of the interaction of light and CO_2 in photosynthesis and growth, other sets of assumptions would give other results in detail. Also, there are significant scaling problems in moving from such static metabolic calculations to the seasonal dynamics in the field. Other processes will have an influence. Review of experimental evidence on the interaction between irradiance and CO_2 concentration in whole plant growth

(Cure 1985; Gates 1985; Kimball 1985; Warrick et al. 1986) is consistent with the above theory that productivity should be CO_2 responsive at all light levels. In all instances C_3 plants at low irradiance responded strongly to elevated CO_2 concentration, sometimes relatively less so than plants under high light, sometimes relatively more so. There may be several reasons why growth of well-illuminated plants is not always relatively more CO_2-sensitive than from low light conditions as the above equations predicted. In part, the exact way the relationship between dark respiration and photosynthesis varies with light and CO_2 may play a role. This will depend on the effect of CO_2 concentration and irradiance on tissue composition and on litterfall and standing live biomass. Source-sink feedbacks in the whole plants may also play a role. Plants growing with abundant illumination, especially determinate ones, may be so well supplied with photoassimilate that sinks grow almost as fast as the prevailing temperature allows, i. e. they are close to substrate saturation. Then further photosynthetic stimulation by elevated CO_2 levels could not lead to much more rapid sink growth and feedbacks would set in suppressing photosynthesis. Such suppression may underpin "photosynthetic acclimation" to high CO_2 (see Sect. 6)

5.1.2 Interaction of CO_2 with Water Regime

Water loss from plants occurs almost entirely through the stomatal pores that open by day thereby allowing CO_2 to diffuse down a concentration gradient to the sites of photosynthetic fixation in the chloroplasts. The price of carbon acquisition is water loss. The evolved strategy of stomatal response to environmental factors appears to be for the aperture to be continuously fine tuned in such a way that water loss is reduced to a degree that nevertheless imposes only marginal restriction to CO_2 diffusion and hence to photosynthesis (Farquhar and Sharkey 1982). This behaviour has most significance where water deficits are constraining growth. Water deficits retard growth (directly or indirectly, Schulze 1986) by restricting stomatal aperture in the very short term and by reducing green leaf area in the longer term (Saab and Sharp 1989) while partitioning relatively more dry matter into roots.

When atmospheric CO_2 concentration is increased, the plant can potentially achieve the same photosynthesis rate with a reduced stomatal aperture. And, in fact, virtually all observations show that high CO_2 concentration reduces stomatal aperture after long-term exposure to high CO_2 concentration, as well as in the short term. Doubling the CO_2 concentration generally immediately reduces stomatal conductance by about 40% (Morison 1985), regardless of the stomatal conductance in normal air. With longer term exposure the magnitude of the response to high CO_2 may be attenuated, the impact on transpiration being offset by increased leaf temperature. Kimball and Idso (1983) concluded from a survey of data in the literature that doubling CO_2 concentration may reduce transpiration by 34% under experimental conditions.

The consequential reduced transpiration rate, under elevated CO_2, reduces any plant-water stress: this tends to permit stomatal aperture to increase again somewhat and leaf expansion to accelerate. This acceleration of leaf expansion, owing to relieved water stress, would be supported by the increased photosynthate supply deriving from the CO_2 effect on photosynthesis. The net effect of these primary effects and series of concurrent feedbacks is that under elevated CO_2 levels, water-restricted plants grow faster, develop greater leaf surfaces, and have reduced stomatal conductance and transpiration per unit leaf surface, though the latter is partially offset by the consequence of warmer leaves (Morison and Gifford 1984a, b). According to studies in controlled environments, the larger the water deficit the greater the relative enhancement to growth owing to elevated CO_2 levels both in C_3 species (Gifford 1979a, b) and C_4 species (Gifford and Morison 1985). The combined effect of reduced transpiration per unit area and increased leaf area can be little or no change in transpiration per plant (Morison and Gifford 1984a), but the efficiency of water use in the formation of plant dry matter is increased in both C_3 and C_4 species when plants are grown on restricted water supplies (Morison and Gifford 1984b).

There is discussion as to whether or not the rate of evapotranspiration at the large scale of regions and continents has changed significantly as atmospheric CO_2 has increased, given the limitations to water vapour exchange imposed by the planetary boundary layer, and given no change in temperature or radiation regime (Jarvis and McNaughton 1986). Since, as mentioned above, water-restricted plants themselves respond to elevated CO_2 by tending to maintain transpiration per plant, while growing faster, it seems unlikely (but not proven) that phenomena in the planetary boundary layer would invalidate conclusions about the CO_2 effect on increasing water-use efficiency that have been drawn from experiments in controlled environments.

So as far as our understanding goes to date, we expect that for communities that are heavily limited in water supply, annual productivity will increase relatively more under elevated CO_2 than for communities which are not water-limited. Productivity of hot, dry desert communities may be particularly increased by the increasing CO_2 concentration, all else being equal.

5.1.3 Interaction of CO_2 with Temperature

Three views have been expressed on the interaction of CO_2 and temperature in plant growth. Based on leaf photosynthetic response to CO_2, it was suggested that the CO_2 responsiveness of productivity would increase progressively with increasing temperature above the optimum (Gifford 1980). This view was supported by a literature survey (Cure 1985) and experimental results subsequent to the review (e.g. Idso et al. 1987).

On the other hand, another literature survey (Kimball 1986) concluded that there was no consistent change in the CO_2 sensitivity of growth as temperature during growth was varied.

A third view (Acock and Allen 1985) is closer to the classic law of the limiting factors concept, namely that plant growth is CO_2-sensitive only at optimal temperatures, either suboptimal or supra-optimal temperatures, eliminating a response to CO_2. This view too has some experimental support (for soybean, Hofstra and Hesketh 1975; tomato, Hand and Postlethwaite 1971; Calvert 1972).

Although adequate growth data on the $CO_2 \times$ temperature interaction are limited because of the high requirement of such studies for scarce controlled environment facilities, there are several theoretical reasons to expect progressively increasing CO_2 responsiveness with increasing temperatures. These reasons concern (1) the photosynthesis/photorespiratory balance; (2) the net photosynthesis/dark respiration balance; and (3) the source/sink balance.

5.1.3.1 Photosynthesis/Photorespiration Balance

The values of kinetic constants of the Rubisco enzyme are temperature-sensitive. These sensitivities have been measured in vitro (Jordan and Ogren 1984) and can be applied in Eqs. (2) and (4). Also, the temperature dependence of gas solubility in water differs for O_2 and CO_2: oxygen loses solubility with increasing temperature less rapidly than does carbon dioxide. The combined effects of the temperature sensitivities of the kinetic constants (expressed with respect to dissolved gas concentration) and the solubilities are such that, for C_3 species, oxygenation increases faster than carboxylation with increasing temperature. Thus, competitive inhibition of oxygenation (i.e. of photorespiration) by elevated CO_2 is stronger at higher temperatures and, consequently, the net photosynthesis rate should be more strongly enhanced by high CO_2 at high temperature. This is illustrated in Fig. 2. The theoretical prediction of a monotonic increase in CO_2 sensitivity of leaf net photosynthesis with increasing temperature is supported by experimental measurements on leaves. For example, Osmond et al. (1980) obtained identical CO_2 sensitivity of light-saturated leaf net photosynthesis as a function of measurement temperature for two species. One species, *Atriplex glabriuscula*, had been grown at 16 °C. The other, *Larrea divaricata*, had been grown at 45 °C. Increasing the CO_2 concentration from about 330 to 1000 ppm increased the net photosynthesis rate of both species by 19, 80 and 194% at measurement temperatures of 16, 30 and 45 °C respectively.

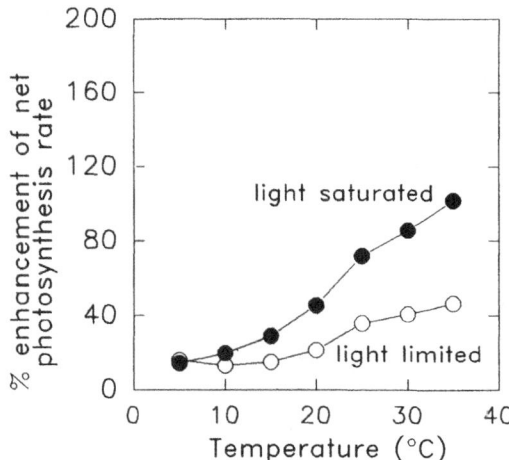

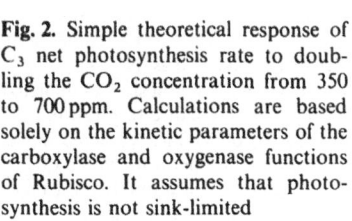

Fig. 2. Simple theoretical response of C_3 net photosynthesis rate to doubling the CO_2 concentration from 350 to 700 ppm. Calculations are based solely on the kinetic parameters of the carboxylase and oxygenase functions of Rubisco. It assumes that photosynthesis is not sink-limited

5.1.3.2 Net Photosynthesis/Dark Respiration Balance

Based mostly on relatively short-term experiments, maintenance respiration is observed to be highly temperature-sensitive, fitting the classical Arrhenius equation (McCree 1974; Amthor 1984). Growth respiration per unit of growth, on the other hand, is considered temperature-independent being dependent mainly on the chemical composition of new growth (Penning de Vries et al. 1974). The rate of light-limited net photosynthesis is relatively insensitive to temperature, because quantum efficiency is temperature-insensitive (Johnson and Thornley 1985), although it often decreases above about 25 °C owing to the increase of photorespiration. Thus, at high temperatures, the negative contribution of respiration to G in Eq. (6) and the impact of a CO_2-driven increase in P_n is more significant than at lower temperatures. This is illustrated in Fig. 3, assuming maintenance respiration is 5% of Pn at 5 °C and has a Q_{10} of 2. The model used for Fig. 3 assumes that growth respiration is 20% of leaf net photosynthesis at all temperatures. Under both light-limited and light-saturated conditions, the percent enhancement of growth rate owing to CO_2 enrichment increases strongly with temperature. This is illustrative of trends only, since a full treatment would require a means of specifying the effect of temperature on standing live biomass.

5.1.3.3 The Source/Sink Balance

The growth rate of biological sinks is highly positively temperature-dependent (Johnson and Thornley 1985). Furthermore, the developmental rate is slow at low temperatures. In many environments, canopy photosynthesis is not very tempera-ture-dependent when measured in short-term experiments. This is because light-limited photosynthesis at 350 ppm CO_2 is only weakly temperature-dependent. Thus, growth is more restricted by low temperature than is the photosynthate supply. That is, growth is determined by sink metabolism, and photosynthesis may be inhibited secondarily by complex feedback regulation (Stitt and Quick 1989), while

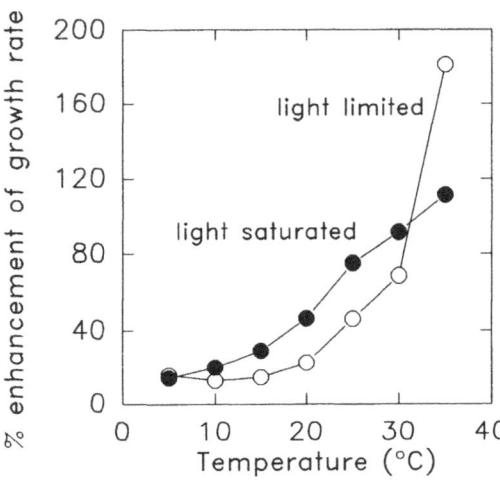

Fig. 3. Simple theoretical response of C_3 plant growth to doubling CO_2 concentration from 350 to 700 ppm. Growth is modelled as $(NP - R_m - R_g)$, where net photosynthesis is as depicted in Fig. 2, maintenance respiration (R_m) is set at 5% of NP at 5 °C increasing as a function of temperature with a Q_{10} of 2, and growth respiration (R_g) consumes 20% of NP at all temperatures

mobilizable storage carbohydrates may accumulate in leaves and stems. Under this circumstance one can expect the potential stimulatory effect of elevated CO_2 levels to be counteracted by metabolic feedback mechanisms that inhibit photosynthesis, i.e. photosynthetic "acclimation" to high CO_2 occurs. Complete adjustment of photosynthesis (and no growth response to CO_2) was reported for the Alaskan tussock tundra sedge, *Eriophorum vaginatum*, growing in situ at 68 °N and 760-m elevation, growing under elevated CO_2 and ambient temperature (Tissue and Oechel 1987). However, in that environment, there is the question of extreme nutrient supply and root growth limitations on water-logged peat soils; in addition, the inherently slow growth of the species might be a factor in the non-responsiveness to CO_2 (Oberbauer et al. 1986).

At high temperatures one expects the converse; high potential sink growth rates may not normally be satiated by photosynthate production, so high potential for CO_2 stimulation of photosynthesis at high temperature would not be counteracted by feedbacks. Thus, at high temperatures one would expect growth to be particularly CO_2-dependent. This has been found at the extreme temperatures used to rid young grapevine cuttings of virus diseases by prolonged growth at 37–40 °C. High CO_2 concentrations more than doubled growth of such plants (Kriedemann et al. 1976). Similarly, cabbage and Chinese cabbage grown at 29–34 °C for 35 days exhibited six-fold larger total plant dry weight when grown under 1000 ppm CO_2 (J.J. Walcott and P.C. Kriedemann 1977, unpubl.).

Nevertheless, data on the CO_2 × temperature interaction are scant and conflicting. There are even reports suggesting that below a certain temperature (e.g. 12 °C for carrot and radish) growth is suppressed by an elevated CO_2 concentration (Idso and Kimball 1989). Although no mechanism has been suggested to explain such a result, it does indicate that much clarification is needed on this topic. Since CO_2 and temperature are the two main "global change" variables, it is important to understand and quantify their interaction.

5.1.4 Interaction of CO_2 with Nutrients

The success of ecosystem primary production models based on radiation (or temperature) and effective water supply (see Sect. 5.1) indicates that, on a broad, long-term global perspective, variation in nutrient supply is not the prime determinant of variations in the primary productivity of natural ecosystems. Nevertheless, there is much scatter of data around the relationships based on radiation and water. Such scatter probably relates in part to soil conditions, especially nutrient supply. Of all the essential mineral nutrients, nitrogen stands out by far as the one to which vegetation, or certainly domesticated crops, are most responsive (Kirby 1981). There is, however, some doubt expressed that native, especially late successional, vegetation which is adapted to nutritionally impoverished sites is necessarily nutrient-limited. The time scale consideration (Sect. 2) enters here, as does the issue of whether humans are removing products, and therefore nutrients, from the ecosystem. There has been little research into the interactions between CO_2 and mineral nutrients in productivity, so I shall confine discussion here to the CO_2 × N interaction which has received the most attention.

Although the widespread short-term response of vegetation growth to artificial additions of N-nutrients to the soil may be regarded as indicating that vegetation productivity is N-limited, in another longer term sense it may not so be N-limited. For the biosphere, the dry matter, which contains about 1% N and 45% C by weight, is immersed in an atmosphere that is 76% N and only 0.053% CO_2 by weight – a stark contrast. Nitrogen-fixing microorganisms are virtually omnipresent on or in plants, litter and soils (Granhall 1981). Most soil and plant organic nitrogen was originally fixed biologically from atmospheric nitrogen. Furthermore, most soils contain huge quantities of organic nitrogen relative to the amount taken up by vegetation each year. The turnover time for soil organic nitrogen is several centuries (Rosswall 1976). In short, roots have potential access to huge quantities of gaseous and combined nitrogen be it largely unavailable in any one season. Its availability either by mineralization or by fixation is dependent in part on the availability of easily oxidized, reduced organic compounds as an energy source for microorganisms. The nitrogen and carbon cycles are tightly linked by the need of soil microorganisms for energy (McGill and Cole 1981). The rate of nitrogen fixation by an ecosystem

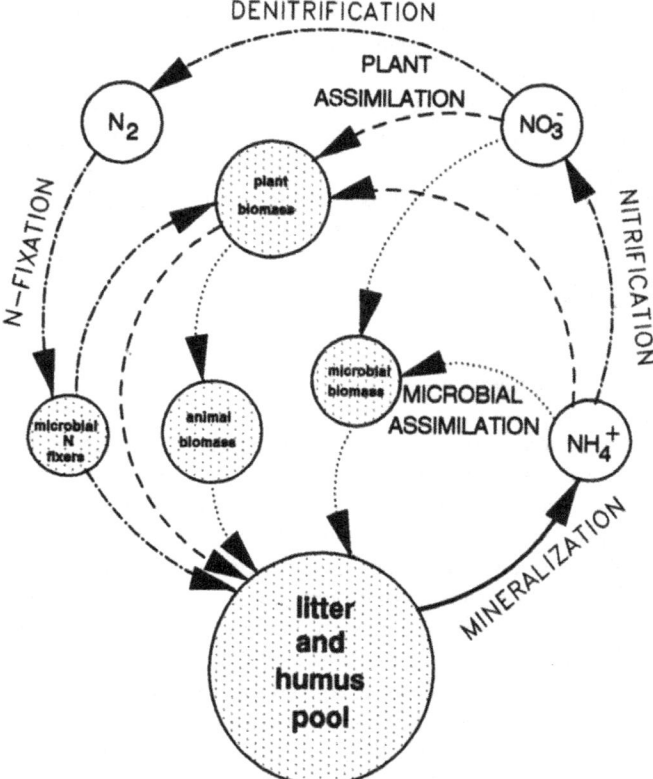

Fig. 4. Simplified "universal nitrogen cycle" (after Jansson 1981). *Dotted circles* depict pools involving sequestered carbon. Three cycles are distinguished: the elemental N-cycle—·—·—·; the autotrophic N-cycle————; and the heterotrophic N-cycle........

is inversely related to its N-content (Woodmanse et al. 1981). The early stages of soil formation involve dominance by N-fixing organisms which give way to non-fixers as the soil N-status builds up (Granhall 1981).

Since the energy supply for bacterial dinitrogen fixation derives from the oxidation of organic matter (Söderlund and Svensson 1976), one can hypothesize that nitrogen fixation may be carbon substrate colimited. Each depends on the other. CO_2 reduction by plants is paced, inter alia, by the seasonal supply of available N: the fixation of atmospheric N is dependent, inter alia, on the supply of reduced carbon (usable energy). According to that concept, artificially increasing the concentration of either available N or atmospheric CO_2 would, over time, increase the fixation of the other as long as other nutrients become available or the C: nutrient ratios widens. Evaluation of the reality of that notion is confounded by the presence of the huge buffers of reduced carbon and reduced nitrogen in the soil which are undergoing continuous turnover and chemical transformation (see Fig. 4; Jansson 1981). Associated mineralization to NH_4^+ and NO_3^- supplies most of the soluble N taken up by plants for growth.

The question, then, is how will the stimulation of plant photosynthesis by elevated atmospheric CO_2 affect the flows and pools in Fig. 4? The question is difficult to answer because we have only sketchy quantification of Fig. 4 on both field and global scales. Also, the answer may differ according to the time scale (from annual to millenial) on which it is addressed. Only a few elements of this complex question can be tackled here.

5.1.4.1 Does N-Limited Plant Growth Respond to Elevated CO_2 Concentration?

For the hypothesis to be sustained that the N-content of an ecosystem will track its carbon content, which will be positively correlated to the atmospheric CO_2 concentration, it is necessary that nitrogen-limited productivity be positively CO_2-sensitive (i.e. colimited).

In some experiments in which potted plants were given regular supplies of nutrient containing various concentrations of N, the percent response of continuous growth to CO_2 enrichment was essentially unaffected by the N-concentration down to highly growth-restrictive levels. This was found in crop species (cotton and maize, Wong 1979) and a wild species (cocklebur, Hocking and Meyer 1985). In other studies the CO_2 sensitivity of growth was less at lower N-supplies (e.g. in several crop species, Goudriaan and de Ruiter 1983). In yet others low N-plants had a greater relative enhancement of growth under elevated CO_2 (Peet and Willits 1984). In a compilation of several studies Kimball (1986) noted great variation in the percent responsiveness of growth to CO_2 as a function of N-concentration in the nutrient solution. There was, however, a tendency for relative enhancement of yield by CO_2 enrichment to be somewhat diminished as N-concentration declined.

When N-limited vegetation grows faster under elevated CO_2, the dry matter accumulated must have a higher C:N ratio. That is, the nitrogen use efficiency is increased by CO_2 enrichment much like water and light use efficiencies are increased by CO_2. At the individual plant level it probably arises in part because the photosynthetic machinery of the plant contains a large proportion of the plant's nitrogen content. Elevating the atmospheric CO_2 concentration has an N-sparing effect, enabling the machinery to fix more carbon per unit of N-investment (Gifford 1989). In short-term glasshouse studies with cotton this came about largely (but not wholly) by the proportion of non-structural carbohydrate in the plant dry weight (especially starch) increasing under both abundant and deficient nitrogen nutrition (Wong 1990). Also, for high productivity salt-marsh vegetation, growing presumably with abundant N-supply, and elevated CO_2 levels throughout the season, increased the C:N ratio of the green shoots (Curtis et al. 1989).

For plants having a symbiotic association with N-fixing microorganisms, CO_2 enrichment may enable faster growth without, necessarily, an increase in the C:N ratio of dry matter. This has been demonstrated several times for domesticated soybean (e.g. Hardy and Havelka 1974; Huang et al. 1975; Masuda et al. 1989). It arises because more nitrogen-fixing nodules, commensurate with the increased plant size, are formed rather than via increased nitrogenase activity per unit nodule mass (Carroll, Gifford and Day, unpubl.). A similar finding was reported by Norby (1987) for three non-domesticated woody perennial species (nodulated legume *Robinia pseudoacacia*, and nodulated actinorhizal species *Alnus glutinosa* and *Eleagnus angustifolia*) growing in N-deficient soil. Doubling the CO_2 concentration increased total plant dry weight by 32, 49 and 61% respectively, and nodule masses were increased by larger proportions. In that study, however, the nitrogen contents per plant were not increased by elevated CO_2 and so the C:N ratio increased. Also, for a wild nodulated soybean (*Glycine soja*), elevated CO_2 did not increase the amount of nitrogen fixed (Masuda et al. 1989). In summary, with this restricted data base, it is unclear whether or not symbiotic nitrogen fixation in natural ecosystems will increase. It may be that for legumes and other symbiotic systems, any CO_2-driven growth occurs at a high C:N ratio.

5.1.4.2 Repercussions for Decomposition and the Soil-Vegetation Nitrogen Cycle

An increase in annual growth could lead to an increase in the annually averaged amount of standing biomass. This would be so, even if the increased growth were partitioned entirely into fast-turnover biomass (litterfall) that is fully oxidized within a year. Only if leaves (and new roots) senesced, abscised and decomposed earlier as a result of faster growth might there be no net effect on the annually averaged amount of C sequestered in standing biomass. Results on leaf senescence, however, are mixed. Some studies have found slight acceleration of senescene under CO_2 enrichment (e.g. in cotton, Chang 1974; soybean, Peet 1984). Such acceleration may be associated with the buildup of starch in the leaves (Ehret and Jolliffe 1985). Alternatively, it might be an expression of nitrogen deficiency through exhaustion of the amount available in the pot. In other studies elevated CO_2 had no impact on leaf senescence (e.g. soybean, Havelka et al. 1984), or delayed senescence (e.g. wheat, Kendall et al. 1985; white oak seedlings, Norby et al 1986a; and a salt marsh sedge, *Scirpus olneyi*, Curtis et al. 1989). Generally, these effects were small. Care in interpretation is needed in case the experimental CO_2 source contained a toxic contaminant, like ethylene, that may accelerate senescence (Morison and Gifford 1984c).

Accumulation of CO_2-enhanced growth into long-term standing biomass (wood) and into surface litter and soil organic matter would represent another route whereby the global increase in atmospheric CO_2 concentration might lead to increased C-sequestration into organic form. Would an increased input of a higher C:N organic litter ratio to the soil lead to more sequestered carbon? If the decay parameters remained the same it would do so, averaged through time, because the larger litter input would be reflected at all stages during the decay time course. However, the rate of litter decay is, inter alia, positively related to its N-content (Turner 1977) and so the rate of decay would be diminished for litter from vegetation growing at elevated CO_2 (Strain and Bazzaz 1983). This has two potential consequences. First, it means that carbon sequestration in litter would increase. Second, it may mean that net nitrogen mineralization would be slowed down and, as a consequence, less nitrogen is available for subsequent uptake. This would act as a negative feedback counteracting the CO_2 stimulus to productivity. Though this could limit the degree of the CO_2 effect on increasing the amount of extra carbon sequestered, the net effect of the changes would probably be more carbon held out of the atmosphere in the plant/litter/soil system involving a substantial relative redistribution from standing to fallen organic matter.

Embedded in such negative feedbacks may be countervailing positive feedbacks. For example, an increased flow into the soil of litter having a high C:N ratio, may foster immobilization of nitrogen thereby reducing the potential for N-loss from the ecosystem by leaching, by denitrification to NO_2 or N_2O, or by volatilization as ammonia (Melillo 1981).

The dynamics of a system response to elevated CO_2 under N-limitation is even more complex and difficult to quantify owing to several further considerations. Besides the C:N ratio, other attributes of litter that determine its rate of decay (and hence N-mineralization) are its content of lignin, which retards decomposition, of tannins, which can either retard or not retard decomposition depending on type, and its soluble carbohydrate content. All these attributes appear to be influenced by the CO_2 content of the atmosphere during plant growth (Norby et al. 1986). Furthermore, it is conceivable that the enhanced root development (e.g. Norby et al. 1986a), and perhaps mycorrhizal association (Lamborg et al. 1983), for vegetation at high CO_2 may enable plants to acquire more of the large pool of cycling soil N. Microorganisms and plant roots compete for the soluble N constantly being produced by gross mineralization. Typically, more than half the mineral N formed is taken up by microorganisms (Rosswall 1976). So if root and mycorrhizal activity increases under elevated CO_2, the vegetation may compete better for the small pool of soluble N (Norby et al. 1984). However, when that idea was examined it did not occur (Norby et al. 1986a).

6 Photosynthetic Acclimation to Elevated CO_2 Levels

Reference was made at the end of Sect. 5.1.1 to "photosynthetic acclimation" that can occur when plants are continuously exposed to high CO_2 concentrations. This not particularly appropriate phrase refers to the phenomenon where the instantaneous net photosynthesis rate per unit area of leaf measured at a standard CO_2

concentration, e.g. either the control or the elevated CO_2 level, is different for leaves from plants grown at the elevated CO_2 than from plants grown in normal atmosphere. That is, where the whole instantaneous CO_2 response curve of photosynthesis per unit leaf area has shifted for plants grown in high CO_2 compared with the curve for plants grown in normal CO_2 concentration. Tissue and Oechel (1987) used the term "homeostatic adjustment" which carries a dynamic regulatory connotation. However, since we are not wholly sure of the basis of the change, it is not yet possible to choose the most apt term. Certainly, there is not always a downward adjustment in the photosynthetic rate per unit area. Sometimes no change was observed, as Gifford (1977) found in the flag leaf of wheat plants grown throughout under elevated CO_2, Jones et al. (1985) found for soybean canopies, Radin et al. (1987) noted for cotton, and Woodrow and Grodzinski (1987) for tomato. In other work, the leaf photosynthetic adjustment to continuous high CO_2 has been upwards, as Bishop and Whittingham (1968) and Hicklenton and Jolliffe (1978) observed for tomato, Conroy et al. (1988) for *Pinus radiata*, Campbell et al. (1988) for soybean and Sage et al. (1989) for potato. The majority of such reports, however, refer to downward adjustment of the leaf photosynthesis rate. The degree of adjustment varies and can even be to such an extent that the adjusted rate measured in elevated CO_2 is equal to or even slightly less than the rate for control plants measured in normal air (e. g. for tobacco, Raper and Peedin 1978; for *Eriophorum vaginatum*, Tissue and Oechel 1987; for the cabbage *Brassica oleracea*, Sage et al. 1989). Various hypotheses have been proposed to explain downward adjustment.

1 Stomatal Conductance. Imai and Murata (1978), working with rice and maize, concluded that downward net photosynthetic adjustment was not related to CO_2 effects on change in dark respiration, photorespiration or mesophyll photosynthetic properties, but was attributable solely to an increase in stomatal resistance to CO_2 diffusion.

2 Mutual Shading. Part of the growth stimulatory effect of elevated CO_2 is a larger leaf area which leads to more mutual shading (Poorter et al. 1988). So, for studies in which the average leaf net photosynthesis rate is estimated via growth analysis or by whole-plant CO_2 exchange, the average irradiance for the leaf area of the plant is reduced when the area is high, thereby reducing the average photosynthesis rate. Furthermore, leaves acclimate to low irradiance by reducing their photosynthetic capacity. Where downward adjustment of photosynthesis follows a prior boost to leaf area development, the leaf area effect can sustain a continued growth response to CO_2 via more light interception until the canopy intercepts all incidence radiation.

3 Choked by Starch Buildup. Sometimes plants grown in an elevated CO_2 concentration accumulate considerable starch in the chloroplasts of the leaf. It has been postulated that this might damage the chloroplast, thereby reducing photosynthesis (Mauney et al. 1979; Cave et al. 1981; Wulff and Strain 1982; DeLucia et al. 1985; Sasek et al. 1985). This seems to be an extreme situation.

4 End-Product-Inhibition by Soluble Carbohydrates. Usually, but not always, the soluble carbohydrate status of the leaves builds up at the same time as starch under elevated CO_2. It is sometimes postulated that these photosynthetic products reduce photosynthesis by feedback inhibition (e.g. Goudriaan et al. 1984). While this is a more subtle proposal than physical disruption by starch granules, it is not considered metabolically likely in a literal sense.

It is now known that in practice there is a hierarchy of metabolic controls centered in the cytosol of the leaf mesophyll cells that acts to link photosynthesis rate, starch synthesis within the chloroplasts and sucrose synthesis in the cytosol with sucrose demand elsewhere in the plant. The hierarchy of mechanisms (involving inter alia transmembrane exchange of triose phosphate and phosphate between the chloroplast and cytosol, sucrose phosphate synthase, fructose 1,6 phosphatase, and the regulatory molecule fructose 2,6-bisphosphate) serve to enable the rate of sucrose synthesis to increase when the photosynthesis rate increases and to reduce sucrose synthesis when the supply of photosynthate exceeds the demand by the plant (Stitt and Quick 1989). This mechanism can operate through the duration of a day causing leaf photosynthesis to decline with time of exposure to light (Foyer 1988) and would presumably be intensified at an elevated CO_2 level.

Thus, the essence of this counteracting acclimation question seems to be whether the vegetation is capable of initiating new sinks in response to the increased soluble carbohydrate supply at elevated CO_2 levels. To explain the upward adjustment of leaf photosynthesis (see above) under this hypothesis, it is necessary to suggest that the earlier availability of more photoassimilate via CO_2 enrichment led to the establishment of more new sinks (e.g. fruits or roots) than the leaf area could sustain without upward photosynthetic adjustment by the feedback mechanisms mentioned.

There are two situations in which the plant might not be able to initiate sufficient new sinks commensurate with the photosynthetic potential at elevated CO_2 levels. First, some species may be genetically pre-disposed not to develop more sinks. Many species that are adapted to extreme nutrient deficiency are unable to grow faster with better mineral nutrition (Chapin et al. 1986) and perhaps the same concept applies to some species in relation to CO_2 nutrition. For example, several rainforest tree species did not respond to elevated CO_2 up to 700 ppm during their first 111 days of life (Reekie and Bazzaz 1989). These were grown under well-illuminated conditions and it would be pertinent to know how they would respond to CO_2 in the low radiation environment of a rainforest gap (see Sect. 5.1.1). In comparisons of the CO_2 response of several annual herbs that grow naturally together, it was found that some grew faster at elevated CO_2 while others did not (Bazzaz and Carlson 1984; Zangerl and Bazzaz 1984; Bazzaz and Garbutt 1988). When that applies we can expect that the community as a whole will respond to elevated CO_2, non-responders giving way to responders. Second, it may be an over-riding environmental constraint that restricts sink demand. For example, an extreme deficiency of phosphate or of a trace element, water-logging, or low temperature might have its primary effect on sink growth properties rather than on photosynthetic source potential. Sufficiently extreme N-deficiency could act like that, too. In such cases one would expect photosynthetic acclimation to occur and the ecosystem productivity not to respond much to elevated CO_2 concentration.

Overall for the biosphere, if we accept that radiation, rainfall and temperature are the main pacers of productivity at normal ambient CO_2, ecosystem nitrogen content being determined ultimately by the carbon input from vegetation productivity, we can expect compensating acclimation to elevated CO_2 to be of only partial significance to most ecosystem productivity as a whole where shifting competitive relations between species can occur. The most probable exception to this is the tundra owing to the low annual temperature combined with water-logging and mineral deficiencies.

7 Expected Overall Impact of Atmospheric CO_2 Increase on Biosphere Primary Productivity

I have argued the case that annual productivity of vegetation is normally colimited by several environmental factors, including CO_2: if a specific resource limitation in the environment were to dominate, the genotypes that invested more in acquiring that resource at the expense of acquiring the less scarce resources would compete and grow better relative to others. In the course of rock weathering and soil formation, vegetation has mobilized unavailable mineral elements from the lithosphere and nitrogen from the atmosphere and stored them in live and dead organic matter and on soil clays, the turnover of which keeps them steadily available. Ecosystems paid a three-fold price for accumulating their store of in-house minerals and constructing their water-storing sponge – the soil and organic matter. The first price was the depletion of the available source of its primary constituent - carbon – to trace levels in the atmosphere by immobilizing it into the lithosphere as solid calcium carbonate. The second price was to concurrently jeopardize itself by building up photosynthetically toxic oxygen gas from virtually zero (Holland et al. 1986) to 21% of the atmosphere. And the third price was the consequential reduction of global temperatures and precipitation to values that are so low that productivity is restricted and a substantial fraction of the land surface is too frozen or too dry to support appreciable vegetation. Furthermore, the oxygen, as an inhibitor of both CO_2 and N_2 fixation and as a prerequisite for regular burning of the biosphere [affecting about 6–7% of the vegetated area of the world per year (Seiler and Crutzen 1980) and releasing about $3Gt(C)$ yr^{-1} as CO_2 (Olson 1982)], has further restricted the average steady-state size of the biosphere. Seen in that long-term context, it seems reasonable to hypothesize that by alleviating the oxygen inhibition of photosynthesis (via competitive inhibition of photorespiration; oxygen depletion is not a foreseeable possibility) there is a prospect that elevated CO_2 levels would increase the overall biosphere primary productivity. The main evidence offered to support that hypothesis is that the other dominant and most prevalent colimiters of global primary production are radiation, water supply and nitrogen supply. High CO_2 increases light-use efficiency, water-use efficiency and nitrogen-use efficiency, and biological nitrogen fixation is dependent on carbohydrate availability. A possible factor working against this hypothesis is that fire frequency and intensity is dependent not only on atmospheric O_2 concentration but also on fuel buildup. It would be expected that increased fuel accumulation in a high CO_2 world could lead to more oxidation of biomass to CO_2 by wildfires. Nevertheless, although it cannot yet be proved, the thrust of the evidence is in the direction of supporting the notion that the 25% increase in atmospheric CO_2 concentration since industrialization started is likely to be increasing global net primary production relative to what it would be without that atmospheric change.

8 Implications for the Global Carbon Cycle and Atmospheric CO_2 Concentrations

Figure 5 depicts part of the global carbon cycle extricated from the longer C-cycles and from other interrelated biogeochemical cycles. It emphasizes that the three C-pools of atmosphere, terrestrial biosphere and soil carbon are exchanging carbon continuously. From Fig. 5, the overall turnover time of the standing biosphere is about 5 years (600/120) and that of the litter plus soil organic matter about 25 years (1500/60). It must be recognized that these are only average turnover times, as there are components within the biosphere and soil-C that persist for up to thousands of years. Nevertheless, the magnitude of these pools and flows indicates that, on the time scale of modern industrial influence and policy, these three pools are in approximate equilibrium. A change in any one element of the cycle will be almost fully equilibrated with the others within a few decades.

The current average rate of atmospheric CO_2 increase, $1.5 \, \text{ppm yr}^{-1}$, corresponds to an atmospheric carbon increase of $3.1 \, \text{Gt(C) yr}^{-1}$. This is $2.2 \, \text{Gt(C)}$ yr^{-1} less than the amount of CO_2 released from fossil fuel burning. This $2.2 \, \text{Gt}$ must be going into oceanic and terrestrial sinks. The network of monitoring sites for CO_2 content of the atmosphere and surface ocean waters around the world has now accumulated sufficient data to permit tentative analyses of the latitudinal location of CO_2 sources and sinks based on spatial and temporal variations in atmospheric CO_2. From an analysis of 1981 to 1987 AD atmospheric and oceanic CO_2 records, Tans et al. (1990) concluded that the global oceans are a net sink for CO_2 to the extent of only $1 \, \text{Gt(C) yr}^{-1}$ at most. The inference is that the land must currently be a net sink for about $1.2 \, \text{Gt(C) yr}^{-1}$ of fossil fuel CO_2. Further, they concluded that this net sink must be in mid- to northern latitudes of the northern hemisphere. Can the land be acting as a net sink despite the deforestation that is occurring mostly in the tropics?

Tropical deforestation is estimated to be occurring at a rate that leads to a net release of carbon from standing biomass, litter and soil organic matter at a rate of 0.4 to $2.6 \, \text{Gt(C) yr}^{-1}$ (Olson 1982; Detwiler and Hall 1988; Houghton and Woodwell 1989). Combining this range of release with the Tans et al. (1990) net estimate of $1.2 \, \text{Gt(C) yr}^{-1}$ that the land should be accumulating indicates that the remaining land (that was not deforested in the year concerned) should be sequestering 1.6 to $3.8 \, \text{Gt(C) yr}^{-1}$.

There are several possible causes of extra carbon sequestering by the land during the 1980s. There is probably a net reforestation occurring in the north temperate regions (Armentano and Ralston 1980; Johnson and Sharpe 1983). Anthropogenic releases into the biosphere of nitrate, phosphate and sulphate via fertilizer and air-pollutant emissions may be fostering the sequestration of carbon into new standing biomass and soil organic matter (Gifford 1987), especially in the major areas of industrialization in the mid-latitudes of the northern-hemisphere. (This would be despite the highly localized productivity decreasing effects of air pollution.) Indeed, the rate of deposition of anthropogenic nitrate and sulphate onto the ice, as far away from industrialization as South Greenland, is increasing rapidly (Mayewski et al. 1990). Here, we focus on the potential of the CO_2-fertilizing effect to accommodate some of this "missing" carbon.

Assuming that the standing biosphere and soil pools are in approximate dynamic equilibrium with the atmospheric C-pool on a time scale of several decades (Fig. 5), one can ask how much extra carbon would need to accumulate in the biosphere and soil pools to reach equilibrium with one year's worth of atmospheric CO_2 increase of 3.1 Gt(C). Taking the linear hypothesis that the marginal partitioning of a small atmospheric increment of CO_2 will, at equilibrium, be equal to overall partitioning, then the incremental amounts will be 3.1:2.5:6.3 Gt(C) yr^{-1} (i.e. in the ratio 740:600:1500) for the atmosphere:live biosphere:dead biosphere. Thus, according to the linear hypothesis (which should be seen as a maximum limit), now that the atmospheric pool has been increasing steadily at 2–3 Gt(C) yr^{-1} for several decades, the storage into live and dead biomass should be approximately 9 Gt(C) yr^{-1}. This alone greatly exceeds the 1.6–3.8 Gt(C) that needs to be accounted for.

In reality, the marginal partitioning of carbon into the organic pools is unlikely to be as large as the absolute partitioning. The opposite extreme – that the marginal partitioning into the organic pools is zero – when viewed in the perspective of this review, is even more unlikely however. Both of the flows between the atmospheric and biospheric pools (Fig. 5) may be influenced by CO_2 concentration. The net photosynthesis flow is unlikely, on present evidence, to be CO_2-saturated except perhaps in extremely cold, nutrient-deficient situations like the tundra. The opposing autotrophic respiration C-flow is likely to stay essentially proportional to the net photosynthesis flow with the possibility, according to minimal evidence to date, that autotrophic respiration as a proportion of net photosynthesis declines with increasing CO_2 concentration, possibily owing to a higher C:N ratio of the tissues formed.

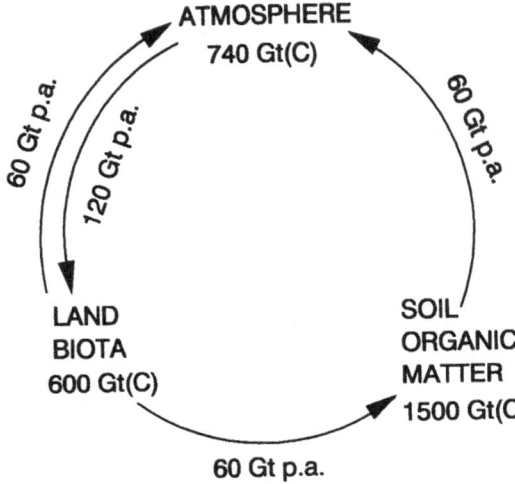

Fig. 5. Approximate carbon pools and flows in the subcycle involving atmosphere, biosphere and soil organic matter. The figures shown are rounded-off values taken from various sources summarized by Bolin (1983) and Hall (1989)

The degree of stimulation of net productivity (and hence litterfall) is far from certain on a global scale but seems likely, on the basis of the evidence reviewed, to be in the range 0.1 to 0.5% for a 1% increase in atmospheric CO_2 concentration [i.e. β, the biotic growth factor of Bacastow and Keeling (1973), $= 0.1 - 0.5$]. To obtain an approximate feeling for the magnitude of the increase of global primary production as a result of the fertilizing effect of 353 ppm compared with the pre-industrial 280 ppm CO_2, we can calculate an upper limit. Assuming, as an upper limit, that for each 1% increase in CO_2 concentration, there has been a 0.5% increase in annual primary productivity of the biosphere, then the 26% increase in atmospheric CO_2 concentration since pre-industrial times would cause a (60 Gt per annum $\times$ 0.26 $\times$ 0.5) 7.8 Gt(C) higher net primary production. This would be an under-estimate of the upper limit to the extent that the CO_2 response curve of productivity is a saturating one and the 0.5 β-factor is derived from calculations and experiments using 330–350 ppm as baseline rather than 280 ppm. Thus, the figure estimated from this "bottom-up" physiological approach is similar to the figure obtained from the "top-down" global carbon pool approach of Fig. 5 using the linear hypothesis. At equilibrium the CO_2-increased productivity would appear as increased above- and below-ground litter production, but whether this would occur from a larger equilibrium standing biomass is not yet possible to ascertain. An increased litterfall of dead biomass having a higher C:N ratio would itself represent an increased biotic carbon storage. The magnitude of this storage would be expected to be increased by a slow-down of decomposition owing to the higher C:N ratio (McGill et al. 1981).

The above-mentioned linear hypothesis of productivity partitioning is, on the basis of the concepts discussed, unlikely. It is more likely that for a 1% step increase in atmospheric CO_2 content, there would be an approximately β% increase in the size of the biospheric and soil plus litter pools, where β is the biotic growth factor for net primary production. Assuming that to be true, then the 9 Gt(C) yr^{-1} potential accumulation, under the linear marginal partitioning hypothesis, would be moderated to a 0.9 to 4.5 Gt(C) yr^{-1} accumulation in the biosphere plus litter and soil carbon pools.

Besides the proposed widening of the C:nutrient ratios of plant and litter organic matter, anthropogenic releases of nitrogen [$\sim$ 90 Mt(N) yr^{-1}], phosphorus [$\sim$ 14 Mt(P) yr^{-1}] and sulphur [$\sim$ 65 Mt(S) yr^{-1}] into the global environment (Peterson and Melillo 1985; Gifford 1987) could support elevated-CO_2 induced carbon storage where these elements become deposited in the landscape. As an unrealistic extreme one can calculate the quantity of carbon potentially sequestered if these anthropogenic mineral nutrient releases were individually converted entirely to standing biomass or to soil organic matter at the prevailing C:nutrient ratios. Taking average element ratios in the live biosphere as C:N:S:P = 790:7.6:3.1:1 (Likens et al. 1981), the above quantities of nutrient would sequester 9, 16, and 11 Gt carbon yr^{-1} for N, S, and P respectively. The equivalent upper limit based on their incorporation into soil organic matter at an average element ratio of C:N:S:P = 54:3:1:2:1 gives a carbon sequestration of 1.6, 2.9 and 0.8 Gt of N, S and P respectively. While biogeochemical cycling ensures that these anthropogenic releases could definitely not be wholly incorporated into live or dead organic matter, the upper limit estimates do suggest that the global eutrophication by concurrent anthropogenic releases of plant-available C, N, P and S could potentially account for the 1.6–3.8 Gt yr^{-1} of missing atmospheric CO_2 indicated by the Tans et al. (1990) analysis.

Finally, we should consider the effect of the increase of about 0.5 °C in the global average temperature over the last century (Kuo et al. 1990). Leith (1972) summarized the empirical relation between net primary productivity of the world's ecosystems and annual average temperature as NPP(t ha^{-1}) = 30/[1 + exp(1.315 − 0.119 × T °C)]. Taking the average annual temperature of the vegetated land surface of the Earth as 18 °C, and an annual primary productivity of 60 Gt(C), a 0.5 °C increase corresponds to about a 1 Gt(C) yr^{-1} productivity increase. Thus, the observed temperature increase may also have increased the productivity of the Earth above that which it otherwise would have been, though not to such an extent as the concurrent CO_2 increase. Warmer conditions tend also to increase the rate of soil organic matter decomposition. The change in the soil organic matter content therefore is a reflection of the balance between the temperature effect on increasing litter deposition and that of increasing the rate of decomposition.

Krapivin and Vilkova (1990) summarized the classical information on soil organic matter content in the top matter as a function of annual average temperature. The relationship varies with annual precipitation. For places with rainfall above 1000 mm yr^{-1}, soil organic matter per m^2 increases with annual temperature up to about 10 °C, above which it remains approximately independent of temperature. Jenny et al. (1949) noted long ago the high organic matter content of tropical soils, relative to some temperate ones, despite the high turnover rates at high temperatures. Thus, for wet tropical areas, global warming would not significantly affect the soil C-store. For areas having less than 1000 mm yr^{-1}, the soil C-store increases with annual average temperature to a maximum of about 10 °C above which it declines with temperature. Thus, in cold, low precipitation areas of the northern forests and tundra it can be expected that global warming, considered alone, would increase the size of the soil C-stores. However, in the hot dry areas, such as deserts and grasslands, global warming would be expected to decrease soil C-amounts, all else being equal. Interestingly, it is in the hot dry areas that the CO_2 fertilizing effect is predicted to be the strongest (see Sect. 5.1.2). Thus, there may be some compensation occurring between the CO_2 and the temperature effects in those areas.

9 Conclusions

An important question is whether the terrestrial biosphere is currently responding to the increasing atmospheric CO_2 concentration and the associated global environmental changes by sequestering more C from the atmosphere. The information and concepts reviewed help in evaluating that question but do not definitively answer it. The magnitude of C-sequestration into live and dead organic matter varies according to the time scale over which it is considered. Since atmospheric CO_2 concentration has been gradually increasing for over a century or more, time scales of response to each annual CO_2 increment covering the full century range are by now operating concurrently. Let us consider, hypothetically in the light of this review, what might happen in response to a single year's increment of atmospheric

CO_2 concentration (now about 1.5 ppm or 0.4% per annum) if that increment occurred all on 1 day of a year.

First, many but possibly not all, C_3 species would initially experience an increase in the photosynthesis rate of about 0.05 to 0.2%. C_4 species would not. All species would experience reduced stomatal conductance by about 0.2%. At least in windy environments, the latter would probably reduce the canopy transpiration rate somewhat. The combined effects on photosynthesis and transpiration lead to an increase in water-use efficiency. The magnitude of these effects would probably be in the upper end of the range in warm and in dry environments. Under such conditions the combined stomatal and photosynthetic effects could lead to a more than 0.2% increase in growth rate in the short term. The consequential increase in the supply of photoassimilates and reduced water stress would lead to an increased rate of leaf expansion, warmer leaves, and a tendency for faster plant growth. Plants that responded to CO_2 more than others could obtain a relatively improved competitive advantage. However, some plants may not be able to generate sufficient new sinks to sustain, for more than a few weeks, the incorporation of the extra photoassimilate into new structural growth. This may be because they are not genetically disposed to grow faster or because other environmental constraints like low temperature, or a restrictive nutrient supply limits the possible growth increment. Then, storage carbohydrate levels would build up and a negative feedback, diminishing the leaf photosynthesis rate, would set in. In ecosystems where the leaf canopies are insufficiently dense to intercept all the incident light owing to drought, it is likely that the negative photosynthetic feedback at the individual leaf level (i.e. photosynthetic acclimation), if it occurs, will develop concurrently with the formation of a larger leaf area. The consequential larger light interception would then sustain a faster growth rate despite the photosynthetic acclimation occurring per unit leaf area. For the most productive ecosystems, which intercept virtually all the incident light, the positive feedback through increased light interception is not possible.

For ecosystems that are mature (i.e. standing biomass, averaged over a big enough area, has reached a stable value from year to year) the rate of new annual growth (net primary production) is matched by the rate of deposition of litter from dead shoot and root parts. During the first few years after such an ecosystem has been exposed to a small step increment in atmospheric CO_2 concentration, there are several possible responses in terms of carbon turnover. For example:

1. Net primary production is increased and this increment is partitioned into all components (leaves, stems, roots, fruits, root litter, shoot litter, etc.) in the same relative proportions as is the base net primary production. Thus, all carbon pools (soil and biomass) would become larger.
2. The increased net primary production is partitioned wholly into the annual litterfall, in which case standing biomass would remain the same, the increased carbon storage being entirely in litter and possibly soil organic matter.
3. The new litterfall would contain the same amount of nitrogen as would have occurred without the CO_2 increment, but more carbon per unit dry matter, owing to a buildup of carbohydrates in the tissues. This would slow its loss to herbivory or its decomposition by saprophytes which thrive on the nitrogen content of live tissues or litter. Such a slow-down in decomposition rate over years would result

in a net carbon storage in litter. At the same time, the more carbonaceous litter could foster particularly the activity of free-living N-fixing microorganisms.

4. Where some aspect of the environment is highly growth restrictive for sinks (e.g. perhaps very low temperatures, trace element deficiency, or waterlogging), photosynthetic acclimation is complete and after a few days of exposure, the leaf photosynthesis rate is the same as before.

From the evidence reviewed, this last response seems unlikely for most ecosystems with the exception of tundra and alpine vegetation which may be too cold for growth to respond permanently to CO_2. More likely, for most vegetation, there will be some combination of the first three responses to a single step increase of CO_2, each pool would reach its new equilibrium carbon content at a different time after the step, according to the turnover time of the pool. For example, the annually averaged carbon sequestered in live leaf would reach its new level after about the first year (at least for deciduous species, which renew leaves completely each year). But this new, higher annual leaf litter drop could contribute to a slowly increasing soil organic matter content for decades or, in cooler climes, with slow decomposition rates, centuries before it was balanced by a matching increase in decomposition rate. It can therefore be expected that parts of the global soil organic matter pool may still be responding to increases in the atmospheric CO_2 content that occurred several decades ago. In the meantime all the subsequent annual increments will be making their contributions additively.

Quantifying the time course of carbon accrual into the global soil profile, following a step increase of atmospheric CO_2, is well beyond our capability at present. Adding further to the problem of quantification is the probability of a widening of the C:N ratio of litter fostering more free-living N-fixation which will, in time via subsequent mineralization, support the enhanced primary productivity potential that the elevated CO_2 offers. So although at first, the rate of N-mineralization may prevent the enhanced productivity potential from being fully expressed, in time there is the possibility that the potential will reach full expression as the nitrogen cycle catches up.

On much longer time scales of millenia and more, it is likely that the CO_2-enhanced, terrestrial biological activity will lead to accelerated rock weathering. This will, via the mechanism described, increase the formation of carbonate shells in the oceans and thereby foster the further sequestration of anthropogenic CO_2 onto the ocean floor where ultimately, after millions of years, it will be subsumed at tectonic plate margins into the magma.

In the meantime, the CO_2 concentration increase is expected to lead to global warming. The global warming is expected to lead to intensification of the hydrologic cycle involving more evaporation from the oceans, higher vapor pressure in the atmosphere and, overall, more rainfall. It is possible, but by no means certain, that it will also involve increased average cloudiness. The combined effects of warming, increased rainfall and increased vapor pressure would strengthen the positive effect of CO_2 on bioproductivity and carbon storage. It is conceivable that increased cloudiness might be associated with the increased atmospheric moisture loading. Any increase in cloudiness, if it led to reduced annual irradiance, would dampen these positive responses. Since none of these climatic changes can yet be adequately quantified, it is still not possible to bring all these threads of global atmospheric change together to predict the magnitude of the positive role of vegetation responses to such change in removing some of the anthropogenic carbon dioxide.

Nevertheless, with the concurrent anthropogenic releases of mineral nutrients and the probably continued increasing temperature due to the greenhouse effect, each tending to cause increased productivity and C-storage, the possibility seems more likely than not that the vegetation of the Earth will continue to sequester a significant portion of the anthropogenic releases assuming, that is, that man does not remove much more of the high-biomass forest cover. Any change in the variability of climate such as decreased temperature variability, increased precipitation variability, and increased frequency and intensity of storms that may accompany global warming (Overpeck et al. 1990) would modify the impact of global atmospheric change on increased bioproductivity but would be unlikely to cancel it.

References

Acock B, Allen LH Jr (1985) Crop responses to elevated carbon dioxide concentrations. In: Strain BR, Cure JD (eds) Direct effects of increasing carbon dioxide on vegetation (DOE/ER-0238) United States Department of Energy, Washington DC, pp 53–97

Amthor JS (1984) The role of maintenance respiration in plant growth. Plant Cell Environ 7:561–569

Armentano TV, Ralston CV (1980) The role of temperate zone forests in the global carbon cycle. Can J For Res 10:53–60

Bacastow R, Keeling CD (1973) Atmospheric carbon dioxide and radiocarbon in the natural carbon cycle. II. Changes from AD 1700 to 2070 as deduced from a geochemical model. In: Woodwell GM, Pecan EV (eds) Carbon and the biosphere. United States Atomic Energy Commission, Washington DC, pp 86–135

Barnola JM, Raynaud D, Korotkevich YS, Lorius C (1987) Vostok ice core provides 160,000-year record of atmospheric CO_2. Nature 329:408–413

Bazzaz FA, Carlson RW (1984) The response of plants to elevated CO_2 I. Competition among an assemblage of annuals at two levels of soil moisture. Oecologia 62:196–198

Bazzaz FA, Garbutt K (1988) The response of annuals in competitive neighborhoods: effects of elevated CO_2. Ecology 69:937–946

Berner RA, Lasaga AC, Garrels RM (1983) The carbonate–silicate geochemical cycle and its effect on atmospheric carbon dioxide over the past 100 million years. Am J Sci 283:641–683

Bishop PM, Whittingham CP (1968) The photosynthesis of tomato plants in a carbon dioxide enriched atmosphere. Photosynthetica 2:31–38

Broecker WS (1973) Factors controlling CO_2 content in the oceans and atmosphere. In: Woodwell GM, Pecan EV (eds) Carbon and the biosphere. U.S. Atomic Energy Commission, Washingtion DC, pp 32–50

Bloom AJ, Chapin FS III, Mooney HA (1985) Resource limitation in plants — an economic analogy. Annu Rev Ecol Syst 16:363–392

Bolin B (1983) The carbon cycle. In: Bolin B, Cook RB (eds) The major biogeochemical cycles and their interactions. (SCOPE 21) John Wiley, Chichester, pp 41–45

Calvert A (1972) Effects of day and night temperatures and carbon dioxide enrichment on yield of glasshouse tomatoes. J Hort Sci 47:231–247

Campbell WJ, Allen LH Jr, Bowes G (1988) Effects of CO_2 concentration on Rubisco activity, amount, and photosynthesis in soybean leaves. Plant Physiol 88:1310–1316

Cave G, Tolley LC, Strain BR (1981) Effect of carbon dioxide enrichment on chlorophyll content, starch content and starch grain structure in *Trifolium subterraneum* leaves. Physiol Plant 51:171–174

Chang CW (1974) Carbon dioxide and senescence in cotton plants. Plant Physiol 55:515–519

Chapin FS III, Vitousek PM, van Cleve K (1986) The nature of nutrient limitation in plant communities. Am Nat 127:48–58

Chapin FS III, Bloom AJ, Field CB, Waring RH (1987) Plant responses to multiple environmental factors. BioScience 37:49–57

Conroy JP, Küppers B, Virgona J, Barlow EWR (1988) The influence of CO_2 enrichment, phosphorus deficiency and water stress on the growth, conductance and water use of Pinus radiata D. Don. Plant Cell Environ 11:91–98

Cure JD (1985) Carbon dioxide doubling responses: a crop survey. In: Strain BR, Cure JD (eds) Direct effects of increasing carbon dioxide on vegetation. DOE/ER-0238, United States Department of Energy, Washington DC, pp 99–116

Curtis PS, Drake BG, Leadley PW, Arp W, Whigham D (1989) Growth and senescence of plant communities exposed to elevated CO_2 concentration on an estuarine marsh. Oecologia 78:20–26

DeLucia EH, Sasek TW, Strain BR (1985) Photosynthetic inhibition after long-term exposure to elevated levels of atmospheric carbon dioxide. Photosynth Res 7:175–184

Detwiler RP, Hall CAS (1988) Tropical forests and the global carbon cycle. Science 239:42–47

Ehret DL, Jolliffe PA (1985) Leaf injury to bean plants grown in carbon dioxide enriched atmospheres. Can J Bot 63:2015–2020

Farquhar GD, Sharkey TD (1982) Stomatal conductance and photosynthesis. Annu Rev Plant Physiol 33:317–345

Foyer CH (1988) Feedback inhibition of photosynthesis through source-sink regulation in leaves. Plant Physiol Biochem 26:483–492

Gates DM (1985) Global biospheric response to increasing atmospheric carbon dioxide concentration. In: Strain BR, Cure JD (eds) Direct effects of increasing carbon dioxide on vegetation. DOE/ER-0238, United States Department of Energy, Washington DC, pp 171–184

Gifford RM (1974a) Comparison of potential photosynthesis, productivity and yield of plant species with differing photosynthetic metabolism. Aust J Plant Physiol 1:107–117

Gifford RM (1974b) Photosynthetic limitations to cereal yield. In: Bieleski RL, Ferguson AR, Cresswell MM (eds) Mechanisms of regulation of plant growth. R Soc NZ Bull 12:887–893

Gifford RM (1977) Growth pattern, CO_2 exchange and dry weight distribution in wheat growing under differing photosynthetic environments. Aust J Plant Physiol 4:99–110

Gifford RM (1979a) Carbon dioxide and plant growth under water and light stress: implications for balancing the global carbon budget. Search 10:316–318

Gifford RM (1979b) Growth and yield of CO_2-enriched wheat under water-limited conditions. Aust J Plant Physiol 6:367–378

Gifford RM (1980) Carbon storage by the biosphere. In: Pearman GI (ed) Carbon dioxide and climate: Australian research. Australian Academy Science, Canberra, pp 167–181

Gifford RM (1982) Global photosynthesis in relation to our food and energy needs. In: Govindjee (ed) Photosynthesis: development, carbon metabolism and plant productivity, vol II. Academic Press, New York, pp 459–495

Gifford RM (1987) Global photosynthesis, atmospheric carbon dioxide and man's requirements. In: Giovannozzi-Sermanni G, Nannipieri P (eds) Proc 7th Int Symp on environmental biogechemistry. CNR-IPRA, Rome, pp 413–443

Gifford RM (1989) The effect of the buildup of atmospheric carbon dioxide on crop productivity. In: Proc 5th Australian agronomy conference, Perth. Australian Society of Agronomy, pp 312–322

Gifford RM, Morison JIL (1985) Photosynthesis, growth and water use of a C_4 grass stand at high CO_2 concentration. Photosynth Res 7:69–76

Gifford RM, Morison JIL, Lambers H (1985) Respiration of crop species under CO_2 enrichment. Physiol Plant 63:351–356

Goudriaan J, de Ruiter HE (1983) Plant growth in response to CO_2 enrichment at two levels of nitrogen and phosphorus supply 1. Dry matter, leaf area and development. Neth J Agric Sci 31:157–169

Goudriaan J, van Laar HH, van Keulen H, Louwerse W (1984) Photosynthesis, CO_2 and plant production. In: Day W, Atkin RK (eds) Wheat growth and modelling. NATO ASI Ser A, Life Science vol 86, pp 107–122

Granhall U (1981) Biological nitrogen fixation in relation to environmental factors and functioning of natural ecosystems. In: Clark PE, Rosswall T (eds) (SCOPE) Ecol Bull (Stockholm) 33:131–143

Hall DO (1989) Carbon flows in the biosphere: present and future. J Geol Soc Lond 146:175–181

Hand DW, Postlethwaite JD (1971) Response to CO_2 enrichment of capillary watered single truss tomatoes at different plant densities and seasons. J Hortic Sci 46:461–470

Hardy RWF, Havelka UD (1974) Photosynthate as a major factor limiting nitrogen fixation by field-grown legumes with emphasis on soybeans. In: Nutman PS (ed) Symbiotic nitrogen fixation in plants. Cambridge University Press, Cambridge, pp 421–439

Havelka UD, Ackerson RC, Boyle MG, Wittenbach VA (1984) CO_2 enrichment effect on soybean physiology. I. Effects of long-term CO_2 exposure. Crop Sci 24:1146–1150

Hendrey GR, Lewin KF, Kolber Z, Kolber D, Lipfert FW, Daum M (1988) Field performance testing of a free-air controlled enrichment (FACE) system for the regulation of carbon dioxide concentrations in a cotton field at Yazoo City, Mississippi. Brookhaven National Laboratory, Department of Applied Science, Report BNL-52194, UC-402 (Environmental Sciences – DOC/OSTI-4500-Interim 3) 44 pp

Hicklenton PR, Jolliffe PA (1978) Effects of greenhouse CO_2 enrichment on the yield and photosynthetic physiology of tomato plants. Can J Plant Sci 58:801–817

Hocking PJ, Meyer CP (1985) Responses of Noogoora Burr (*Xanthium occidentale* Bertol) to nitrogen supply and carbon dioxide enrichment. Ann Bot 55:835–844

Hofstra G, Hesketh JD (1975) The effects of temperature and CO_2 enrichment on photosynthesis in soybean. In: Marcelle R (ed) Environmental and biological control of photosynthesis. Junk, The Hague, pp 71–80

Holland HD, Lazar B, McCaffrey M (1986) Evolution of the atmosphere and the oceans. Nature 320:27–33

Houghton RA, Woodwell GM (1989) Global climate change. Sci Am 260:36–44

Huang C–Y, Boyer JS, Vanderhoef LN (1975) Limitations of acetylene reduction (nitrogen fixation) by photosynthesis in soybean having low water potentials. Plant Physiol 56:228–232

Idso SB, Kimball BA (1989) Growth response of carrot and radish to atmospheric CO_2 enrichment. Environ Exp Bot 29:135–139

Idso SB, Kimball BA, Anderson MG, Mauney JR (1987) Effects of atmospheric CO_2 enrichment and plant growth: the interactive role of air temperature. Agric Ecosyst Environ 20:1–10

Imai K, Murata Y (1978) Effect of carbon dioxide concentration on growth and dry production of crop plants. V. Analysis of after-effect of carbon dioxide treatment on apparent photosynthesis. Jpn J Crop Sci 47:587–595

Jansson SL (1981) Mineralization and immobilization of soil nitrogen by microorganisms: Rapporteur's comments. In: Clark FE, Rosswall T (eds) Terrestrial nitrogen cycles. Processes, ecosystem strategies and management impacts. (SCOPE) Ecol Bull 33:195–200

Jarvis PG, McNaughton KG (1986) Stomatal control of transpiration: scaling up from leaf to region. Adv Ecol Res 15:1–49

Jenny H, Gessel SP, Bingham FT (1949) Comparative study of decomposition rates of organic matter in temperate and tropical regions. Soil Sci 68:419–432

Johnson IR, Thornley JHM (1985) Temperature dependence of plant and crop processes. Ann Bot 55:1–24

Johnson WC, Sharpe DM (1983) Evaluation of merchantable-total biomass conversion ratios used in global carbon budget research. Can J For Res 13:372–383

Jones P, Allen LH Jr, Jones JW, Valle R (1985) Photosynthesis and transpiration responses of soybean canopies to short-and long-term CO_2 treatments. Agron J 77:119–126

Jordan DB, Ogren WL (1984) The CO_2/O_2 specificities of ribulose 1,5-bisphosphate carboxylaseloxygenase. Planta 161:308–313

Kacser H, Burns JA (1973) The control of flux. Symp Soc Exp Biol 27:65–104

Kendall AC, Turner JC, Thomas SM, Keys AJ (1985) Effects of CO_2 enrichment at different irradiances on growth and yield of wheat. II Effects on Kleiber spring wheat treated from anthesis in controlled environments in relation to effects on photosynthesis and photorespiration. J Exp Bot 36:261–273

Kimball BA (1985) Adaptation of vegetation and management practices to a higher carbon dioxide world. In: Strain BR, Cure JD (eds) Direct effects of increasing carbon dioxide on vegetation, DOE/ER-0238 United State Department of Energy, Washington DC pp 185–204

Kimball BA (1986) CO_2 stimulation of growth and yield under environmental restraints. In: Enoch HZ, Kimball BA (eds) Carbon dioxide enrichment of greenhouse crops. II. Physiology, yield, and economics. CRC Press Boca Raton, Florida pp 53–68

Kimball BA, Idso SB (1983) Increasing atmospheric CO_2 effects on crop yield, water use and climate. Agric Water Manage 7:55–72

Kirby EA (1981) Plant growth in relation to nitrogen supply. In: Clark FE, Rosswall T (eds) Terrestrial nitrogen cycles: processes, ecosystem strategies and management impacts. (SCOPE) Ecol Bull (Stockholm) 33:249–267

Kramer PJ (1981) Carbon dioxide concentration, photosynthesis and dry matter production. Bioscience 31:29–33

Krapivin VF, Vilkova LP (1990) Model estimation of excess CO_2 distribution in biosphere structure. Ecol Model 50:57–78

Kriedemann PK, Sward RJ, Downton WJS (1976) Vine response to carbon dioxide enrichment during heat therapy. Aust J Plant Physiol 3:605–618

Kuo C, Lindberg C, Thomson DJ (1990) Coherence established between atmospheric carbon dioxide and global temperature. Nature 343:709–714

Lamborg MR, Hardy RWF, Paul EA (1983) Microbial effects. In: Lemon ER (ed) CO_2 and plants: the response of plants to rising levels of atmospheric carbon dioxide. Westview Press, Boulder, Colorado, pp 131–176

Leith H (1972) Modelling the primary productivity of the world. Nat Resour 8:5–10

Lemon ER (1977) The land's response to more CO_2. In: Anderson NR, Malahoff A (eds) The fate of fossil fuel CO_2 in the oceans. Plenum, New York, pp 97–130

Liebig J Von (1855) Grundsätze der Agrikultur-Chemie mit Rücksicht auf die in England angestellten Untersuchungen. F Viewegund Sohn, Braunschweig, Germany

Likens EG, Bormann FH, Johnson NM (1981) Interactions between major biogeochemical cycles in terrestrial ecosystems. In: Likens GE (ed) Some perspectives of the major biogeochemical cycles. (SCOPE 17) John Wiley, Chichester, pp 93–112

Lovelock JE, Whitfield M (1982) Life span of the biosphere. Nature 296:561–563

Masuda T, Fujita K, Kogure K, Ogata S (1989) Effect of CO_2 enrichment and nitrate application on vegetative growth and nitrogen fixation of wild and cultivated soybean varieties. Soil Sci Plant Nutr 35:357–366

Mattson WJ, Addy ND (1975) Phytophagous insects as regulators of forest primary production. Science 190:515–522

Mauney JR, Guinn G, Fry KE, Hesketh JD (1979) Correlation of photosynthetic carbon dioxide uptake and carbohydrate accumulation in cotton, soybean, sunflower and sorghum. Photosynthetica 13:260–266

Mayewski PA, Lyons WB, Spencer MJ, Twickler MS, Buck CF, Whitlow S (1990) An ice-core record of atmospheric response to anthropogenic sulfate and nitrate. Nature 346:554–556

McCree KJ (1974) Equations for the rate of dark respiration of white clover and grain sorghum as functions of dry weight, photosynthetic rate and temperature. Crop Sci 14:509–514

McGill WB, Cole CV (1981) Comparative aspects of cycling of organic C, N, S and P through organic matter. Geoderma 26:267–286

McGill WB, Hunt HW, Woodmanse RG, Reuss JO (1981) Phoenix, a model of the dynamics of carbon and nitrogen in grassland soils. In: Clark FE, Rosswall T (eds) Terrestrial nitrogen cycles: processes, ecosystem strategies and management impacts. (SCOPE) Ecol Bull 33:49–115

Melillo JM (1981) Nitrogen cycling in deciduous forests. In: Clark FE, Rosswall T (eds) Terrestrial nitrogen cycles: processes, ecosystem strategies and management impacts. (SCOPE) Ecol Bull 33:427–442

Morison JIL (1985) Intercellular CO_2 concentrations and stomatal response to CO_2. In: Zeiger E, Cowan IR, Farquhar GD (eds) Stomatal function. Stanford University Press, Stanford, California, Chap 14

Morison JIL, Gifford RM (1984a) Plant growth and water use with limited water supply in high CO_2 concentration. I. Leaf area, water use and transpiration. Aust J Plant Physiol 11:361–374

Morison JIL, Gifford RM (1984b) Plant growth and water use with limited water supply in high CO_2 concentration. II. Plant dry weight, partitioning and water use efficiency. Aust J Plant Physiol 11:375–384

Morison JIL, Gifford RM (1984c) Ethylene contamination of CO_2 cylinders: effects on plant growth in CO_2 enrichment studies. Plant Physiol 75:275–277

Norby RJ (1987) Nodulation and nitrogenase activity in nitrogen-fixing woody plants stimulated by CO_2 enrichment of the atmosphere. Physiol Plant 71:77–82

Norby RJ, Luxmoore RJ, O'Neill NG, Weller DG (1984) Plant responses to elevated atmospheric CO_2 with emphasis on below-ground processes. Rep ORNL/TM-9426, Oak Ridge National Laboratory, Oak Ridge, Tennessee

Norby RJ, O'Neill EG, Luxmore RJ (1986a) Effects of atmospheric CO_2 enrichment on the growth and mineral nutrition of *Quercus alba* seedlings in nutrient poor soil. Plant Physiol 82:83–89

Norby RJ, Pastor J, Melillo JM (1986b) Carbon–nitrogen interactions in CO_2 enriched white oak: Physiological and long term prespectives. Tree Physiol 2:233–241

Oberbauer SF, Sionit N, Hastings SI, Oechel WC (1986) Effects of CO_2 enrichment and nutrition on growth, photosynthesis, and nutrient concentration of Alaskan tundra plant species. Can J Bot 64:2993–2998

Olson JS (1982) Earth's vegetation and atmospheric carbon dioxide. In: Clark WC (ed) Carbon dioxide review: 1982. Clarendon, Oxford, pp 388–398

Osmond CB, Björkman O, Anderson DJ (1980) Physiological processes in plant ecology: toward a synthesis with Atriplex. Springer, Berlin Heidelberg New York, 468 pp

Overpeck JT, Rind D, Goldberg R (1990) Climate-induced changes in forest disturbance and vegetation. Nature 343:51–53

Pearman GI (1988) Greenhouse gases: evidence for atmospheric changes and anthropogenic causes. In: Pearman GI (ed) Greenhouse: planning for climate change. EJ Brill, Leiden pp 3–21

Peet MM (1984) CO_2 enrichment of soybeans. Effects of leaf/pod ratio.Physiol Plant 60:38–42

Peet MM, Willits DH (1984) CO_2 enrichment of greenhouse tomatoes using a closed-loop heat storage: effects of cultivar and nitrogen. Sci Hortic 24:21–32

Penning de Vries FWT, Brusting AHM, van Laar HH (1974) Products, requirements and efficiency of biosynthetic processes: a quantitative approach. J Theor Biol 45:339–377

Peterson BJ, Melillo JM (1985) The potential storage of carbon caused by eutrophication of the biosphere. Tellus 37B:117–127

Poorter H, Pot S, Lambers H (1988) The effect of an elevated atmospheric CO_2 concentration on growth, photosynthesis and respiration of *Plantago major*. Physiol Plant 73:553–559

Radin JW, Kimball BA, Hendrix DL, Mauney JR (1987) Photosynthesis of cotton plants exposed to elevated levels of carbon dioxide in the field. Photosynth Res 12:191–203

Raper CD, Peedin GF Jr (1978) Photosynthetic rate during steady-state growth as influenced by carbon dioxide concentration. Bot Gaz 139:147–149

Reekie EG, Bazzaz FA (1989) Competition and patterns of resource use among seedlings of five tropical trees grown at ambient and elevated CO_2. Oecologia 79:212–222

Reuveni J, Gale J (1985) The effect of high levels of carbon dioxide on dark respiration and growth of plants. Plant Cell Environ 8:623–628

Rosswall T (1976) The internal nitrogen cycle between microorganisms, vegetation and soil. In: Svensson BH, Söderlund R (eds) Nitrogen, phosphorus and sulphur – global cycles. (SCOPE 7) Ecol Bull (Stockholm) 22:157–168

Saab IN, Sharp RE (1989) Non-hydraulic signals from maize roots in drying soil: inhibition of leaf elongation but not stomatal conductance. Planta 179:466–474

Sage RF, Sharkey TD, Seeman JR (1989) Acclimation of photosynthesis to elevated CO_2 in five C_3 species. Plant Physiol 89:590–596

Sasek TW, DeLucia EH, Strain BC (1985) Reversibility of photosynthetic inhibition in cotton after long-term exposure to elevated CO_2 concentration. Plant Physiol 78:619–622

Schulze ED (1986) Whole plant response to drought. Aust J Plant Physiol 13:127–141

Schwartzman DW, Volk T (1989) Biotic enhancement of weathering and the habitability of earth. Nature 340:475–460

Seiler W, Crutzen PJ (1980) Estimates of gross and net fluxes of carbon between the biosphere and the atmosphere from biomass burning. Clim Change 2:207–247

Sharkey TD (1986) Theoretical and experimental observations on O_2-sensitivity of C_3 photosynthesis. In: Marcelle R, Clijsters H, van Poucke M (eds) Biological control of photosynthesis. Martinus Nijhoff, Dordrecht, pp 115–125

Silsbury JH, Stevens R (1984) Growth efficiency of *Trifolium subterraneum* at high [CO_2]. In: Sybesma C (ed) Advances in photosynthesis research, vol IV. Martinus Nijhoff/W Junk, The Hague, pp IV.2.133–136

Singh G (1988) History of arid land vegetation and climate: a global prespective. Biol Rev 63:159–195

Söderlund BH, Svensson R (1976) The global nitrogen cycle. In: Svensson BH, Söderlund R (eds) Nitrogen, phosphorus and sulphur – global cycles (SCOPE 7) Ecol Bull (Stockholm) 22:23–74

Stitt M, Quick WP (1989) Photosynthetic carbon partitioning: its regulation and possibilites for manipulation. Physiol Plant 77:633–641

Strain BR, Bazzaz FA (1983) Terrestrial plant communities. In: Lemon ER (ed) CO_2 and plants: the response of plants to rising levels of atmospheric carbon dioxide. Westview Press, Boulder, Colorado, pp 177–222

Tans PP, Fung IV, Takahashi T (1990) Observational constraints on the global atmospheric CO_2 budget. Science 247:1431–1438

Tissue DT, Oechel WC (1987) Response of *Eriophorum vaginatum* to elevated CO_2 and temperature in the Alaskan tussock tundra. Ecology 68:401–410

Turner J (1977) Effect of nitrogen availability on nitrogen cycling in a Douglas-fir stand. For Sci 23:307–316

Uchijima Z, Seino H (1985) Agroclimatic evaluation of net primary productivity of natural vegetations I. Chikugo model for evaluating net primary productivity. J Agric meteorol 40:343–352

Uchijima Z, Seino H (1987) Maps of net primary productivity of natural vegetation on continents. National Institute of Agro-Environmental Sciences, Kyushu National Agricultural Station, Kyushu, Japan (in Japanese with English summary), 102 pp

Volk T (1987) Feedbacks between weathering and atmospheric CO_2 over the last 100 million years. Am J Sci 287:763–779

Volk T (1989) Rise of angiosperms as a factor in long-term climatic cooling. Geology 17:107–110

von Caemmerer S, Farquhar GD (1981) Some relationships between the biochemistry of photosynthesis and the gas exchange of leaves. Planta 153:376–387

Walker JCG (1984) How life affects the atmosphere. BioScience 34:486–491

Walker JCG, Hays PB, Kasting JF (1981) A negative feedback mechanism for the long-term stabilization of earth's surface temperature. J Geophys Res 86 (C10): 9776–9782

Warrick RA, Gifford RM, Parry M (1986) CO_2 climate change and agriculture. In: Bolin B, Döös BR, Jäger J, Warrick RA (eds) The greenhouse effect, climatic change, and ecosystems. John Wiley, Chichester, pp 393–474

White TCR (1984) The abundance of invertebrate herbivores in relation to the availability of nitrogen in stressed food plants. Oecologia 63:90–105

Wong SC (1979) Elevated atmospheric partial pressure of CO_2 and plant growth. I. Interactions of nitrogen nutrition and photosynthetic capacity in C_3 and C_4 plants. Oecologia 44:68–74

Wong SC (1990) Elevated atmospheric partial pressure of CO_2 and plant growth. II. Non-structural carbohydrate content in cotton plants and its effect on growth parameters. Photosynth Res 23:171–180

Woodmanse RG, Vallis I, Mott JJ (1981) Grassland nitrogen. In: Clark FE, Rosswall T (eds) Terrestrial nitrogen cycles: processes, ecosystem strategies and management impacts. (SCOPE) Ecol Bull (Stockholm) 33:443–462

Woodrow L, Grodzinski B (1987) Photosynthetic gas exchange, photoassimilate partitioning and development in tomato under CO_2 enrichment. In: Biggins J (ed) Progress in photosynthesis research, vol III. Martinus Nijhoff, Dordrecht, pp III.9.653–656

Wulff R, Strain BR (1982) Effects of carbon dioxide enrichment on growth and photosynthesis in *Desmodium paniculatum*. Can J Bot 60:1084–1091

Zangerl AR, Bazzaz FA (1984) The response of plants to elevated CO_2 II. Competitive interactions between annual plants under varying light and nutrients. Oecologia 62:412–417

Radiative Transfer in Nonhomogeneous Plant Canopies

T. NILSON

Contents

List of Symbols

x, y, z	Cartesian coordinates associated with a plant or plant canopy
X	A point in a plant or in a plant canopy
ξ, η	Coordinates associated with crown envelope projection region
$\Omega = (\theta, \phi)$	Unit vector in the direction determined by the zenith angle θ and azimuth ϕ
Ω_0	Unit vector in a solar direction
$g(\Omega_L)$	Distribution density of foliage elements' normals
$G(\Omega)$	Mean projection of unit foliage area on the plane perpendicular to the direction Ω
$\Gamma(\Omega' \rightarrow \Omega)$	(Area) scattering phase function for the canopy medium
$u(X)$	Foliage area volume density at the point X
H	Plant or canopy height
h	Crown length

R	Crown radius
$S(z, \Omega)$	Crown envelope projection area on a horizontal plane at the height z in the direction Ω
$V(z)$	Volume of the crown envelope chopped at the height z
$T(z, \Omega)$	Crown envelope projection region
$f_\Omega(\xi, \eta)$	Probability density of projected foliage elements
c_Ω	Total projected foliage area
$\alpha_1(X, \Omega)$	Binary random gap function
$a_1(z, \Omega)$	Average proportion of gaps in a single plant crown
a_s	Direct solar radiation penetration coefficient
a_D	Diffuse sky radiation penetration coefficient
$a(X, \Omega)$	Average proportion of gaps in the plant canopy at the point X in the direction Ω
$Q_1(X; \Omega, \Omega')$	Bidirectional gap probability in a single plant
$Q(X; \Omega, \Omega')$	Bidirectional gap probability in the plant canopy
l	Ray path length in the crown envelope
τ	Extinction coefficient per unit path length in a crown
r	Reflection coefficient of a foliage element
t	Transmission coefficient of a foliage element
P	Various probabilities
ρ	Canopy reflectance factor

1 Introduction

The transfer of solar radiation in plant canopies has been explored both theoretically and experimentally during the last three decades. Typically, radiation transfer in plant canopies is examined in terms of three problems: (1) Theoretical modeling of the phyotosynthetic productivity in plant stands and production ecology; (2) heat and mass transfer in vegetation as a boundary layer to the earth's atmosphere; (3) remote sensing of agricultural crops, forests and other types of man-created or natural vegetation.

In photosynthetic studies, the main emphasis has been on the spatial and temporal distribution of absorbed photosynthetically active radiation (PAR, 400–700 nm) in the plant canopy. It is also essential to consider integral shortwave radiation as a contributor to the heat budget when estimating the effect of leaf temperature on physiological processes, such as photosynthesis, respiration, transpiration and growth. Similarly, atmospheric boundary layer problems require the total amount of radiative energy absorbed and reflected by vegetation (e.g. canopy albedo). For remote sensing purposes mainly the directional canopy reflectances in several narrow spectral regions are required. However, all the above problems are related to radiative transfer in vegetation.

Since the publication of the pioneer work by Monsi and Saeki (1953) over 70 theoretical models of canopy radiation regime have been proposed. Principally, most of these models differ only in a few details. Here, the reader is referred to some review papers: Lemeur and Blad (1974), Norman (1975), books by Ross (1981) and Myneni and Ross (1991) and the recent extensive reviews by Goel (1988) and Myneni et al. (1989). I shall present my personal understanding of the radiative transfer problem formed by these papers. In particular, I have been influenced by Ross (1964, 1972, 1981), Charles-Edwards and Thornley (1973), Allen (1974), Norman (1975), Norman and Jarvis (1975), Nilson (1977), Nilson et al. (1977), Mann et al. (1977, 1979), Vygodskaya (1981), Kimes and Kirchner (1982), Norman and Welles (1983), Anisimov and Menzhulin (1983), Oker-Blom (1984, 1986), Kuusk (1985), and many others.

A prevailing majority of theoretical papers on modeling of radiation transfer begin with nearly the same phrase: consider a plant canopy homogeneous in the horizontal. Vertical homogeneity is declared seldom. Horizontal homogeneity is extremely rare in real plant canopies, including cultivated crops. Row structure is often noticeable even in a closed cereal canopy.

The main reasons for horizontal nonhomogeneity are: (1) foliage elements of a plant are located within a certain envelope; as a result, in sparser plant stands there are considerable gaps between individual plants; (2) the volume density of plant material inside the envelope is variable; (3) the density of plants and species composition vary regularly or randomly; (4) the size and foliage area volume densities are different for different plants.

Clumping or clustering is a typical case of nonhomogeneity. Here, a sparse coniferous forest serves as a good example. Three or four different clustering levels may be distinguished (Norman and Jarvis 1975; Oker-Blom 1986): shoot level, needles are grouped into shoots; branch level, shoots are grouped into branches; whorl level, branches are grouped into whorls; crown level, whorls are grouped into crowns.

Analogous to homogeneous canopies, the total radiation field within those which are nonhomogeneous can be reasonably expressed as a sum of three components (Niilisk et al. 1970):

1. Nonintercepted part of direct solar radiation;
2. Nonintercepted part of diffuse sky radiation;
3. Radiation scattered at least once inside the plant canopy.

The free path length of photons in the canopy is determined by the presence of gaps in the foliage. Thus, components (1) and (2) can be described by canopy structural parameters only, while component (3) needs the consideration of optical parameters as well.

2 Vegetation Canopy as an Array of Individual Subcanopies

In order to obtain more realistic canopy radiation models, in contrast of the concept of horizontally homogeneous leaf canopies, the concepts of foliage envelopes (Charles-Edwards and Thornley 1973) or subcanopies (Norman and Welles 1983; Whitfield 1986) should be accepted. Moreover, for conifers it seems reasonable to introduce individual foliage envelopes for each clustering level: crown, whorl or branch and shoot envelopes. The plant stand structure is then simulated in a realistic

manner for groups of envelopes on each clustering level specified in some arrangement, e.g. certain random or systematic distribution patterns. Such an approach
permits the investigator to consider separately the influence of between-plant shading
and within-plant shading on radiation penetration, which is quite essential in an
ecological sense. At a smaller level of the structural hierarchy, within-shoot shading
may be distinguished from between-shoot shading for conifers (Oker-Blom and
Kellomäki 1983; Oker-Blom 1986).

Hence, in the treatment below a plant canopy is assumed to be comprised of
individual plants of different sizes and shapes with a given horizontal pattern for
individuals. Clearly, a plant stand is not a mechanical composition of fairly independent plants. Plants in the community affect each other markedly. Here, one of the
most significant but poorly studied problems is how to describe this mutual interaction quantitatively.

In some practical cases, the rough concept of a homogeneous leaf canopy seems
acceptable, however, as a convenient approximation of the real situation.

3 Isolated Plant. Modeling of Plant Structure and Radiation Regime

The structure of an individual plant as the main constituent of the plant canopy
plays an important role in forming the radiation conditions inside the plant canopy.
Theoretical considerations of radiation penetration and interception should start
with the choice of a certain structural plant model. To meet the requirements of
radiation modeling a plant model must include the following characteristics:
(1) dimensions and shape of the foliage envelope; (2) total foliage area and its 3-D
distribution inside the crown envelope, if possible, separately for each plant organ
[leaves, branches, stem(s), reproductive organs]; (3) foliage inclination and orientation distribution.

A plant model should also consider the most significant, genetically determined
structural features, such as phyllotaxis, internode relations, perhaps leaf size and
shape, and even the presence of heliotropic movement in leaves. At the same time
the model should be able to reproduce the changes in the plant structure introduced
by the interaction with other plants in the community and those caused by environmental factors. Recently, an interesting and promising method of simulating plant
structures has been proposed by Borel-Donohue (1988). He used fractal tree models
to recreate the dimensions, positions and orientations of branches and leaves.

For simplicity, consider a plant consisting of a vertical stem and a crown. Call
the whole, above-ground part of the plant, except for the bottom part of the stem
uncovered by branches or leaves, a plant crown. Other plant forms, e.g. those having
several or inclined stems, can be treated similarly.

Here, it is convenient to introduce a Cartesian coordinate system (x, y, z)
associated with the plant. Let the origin of the coordinates coincide with the origin
of the plant on the ground. Let the horizontal coordinate x be directed towards
the longer axis of plant symmetry, provided it exists (see Fig. 1). Directions will be
given as the respective unit vectors $\Omega = (\theta, \phi)$, where θ is the zenith angle and ϕ is
the azimuth of the direction Ω.

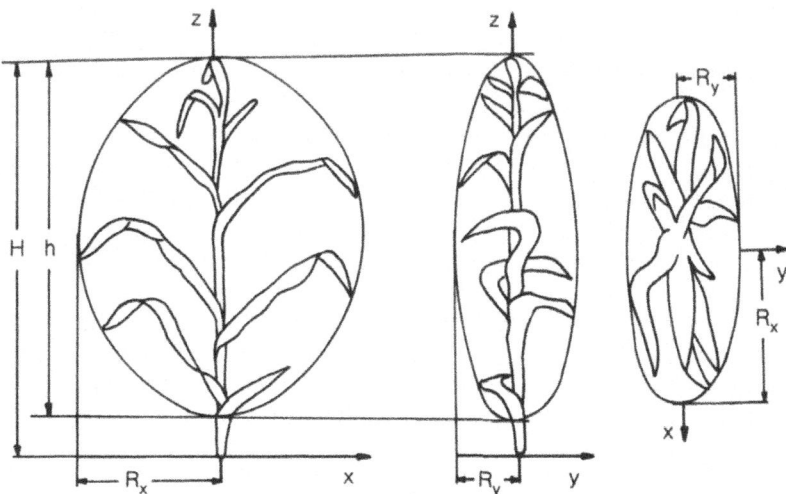

Fig. 1. A maize plant silhouette together with the coordinate system (x, y, z) associated with the plant and a crown envelope shown in three perpendicular projections

3.1 Crown Envelope

A great number of plant crown models have been proposed using certain geometrical figures, such as ellipsoids, cones, cylinders, etc. The main aim has been the determination of the crown volume or crown surface area. The choice of the well-known geometrical figures has been motivated by the relative simpleness of the respective analytical formulas for the crown volume and area. Evidently, a more adequate treatment needs the introduction of more complex geometrical figures. Here, the investigator must decide how to determine the crown envelope.

Define the crown envelope as an ideal convex and smooth enveloping surface containing the whole range of crown foliage elements. Certainly, this determination is somewhat arbitrary. One possible way of defining quantitatively the crown envelope for each individual plant is the so-called rolling disc method or, for a 3-D case, the rolling sphere method. The idea is rather simple: a disc of a certain radius is rolled down along the crown periphery so that it only touches the outermost foliage elements. The trajectory formed by the disc edge may be treated as the crown envelope. The greater the disc radius, the less the small details of the foliage structure are followed and the smoother is the canopy envelope obtained. Evidently, for each species an appropriate disc radius should be determined. Using this method, the crown envelope must be calculated on the computer, provided the investigator has succeeded in inserting the crown structure in the computer memory. Here, a photographic method or a method of silhouettes (Loomis et al. 1967) may be used. It seems relatively easy to apply the rolling disc method on computer-simulated plants, e.g. on those obtained by the algorithms based on fractal properties (Borel-Donohue 1988).

As a result of applying the rolling disc method to each individual plant, the crown envelope is a random function. Having determined in such a manner the envelopes for a set of plants of nearly the same size, some statistical envelope characteristics may be estimated. First, we must determine the average crown envelope. Other statistical characteristics may also appear to be of interest. Knowing the vertical course of the variance we can decide how well the mean crown envelope matches the real situation and whether it affects radiation calculations. In a more sophisticated treatment, an analysis of the correlation function of the random crown envelope is required. It is likely that such a statistical analysis will make it possible to distinguish between the variability in the crown envelope shape caused by genetic factors and the community-related variability. However, this remains a topic of further research.

In order to derive mathematical expressions or computer algorithms for the projection area of the crown envelope, and for the volume and path length of rays in various view directions, a mathematical description of the crown shape is needed. Modeling of the crown envelope as a body of rotation needs an equation for the rotating curve.

Using a photographic method, Oker-Blom et al. (1986) determined the shape of the crown envelope for some young Scots pines. They modeled crown envelopes as bodies of rotation composed of three parts: a cone at the top, a cylinder in the middle and an intersected sphere at the lowest part of the crown. The length of the conical upper part was approximately half of the total crown length.

The choice of an appropriate model for the crown shape depends definitely on the phyllotaxis of the plant. For instance, plants having a genetic spiral angle $\alpha = 180°$ may often be approximated as ellipsoides with considerably different horizontal axes. Nilson et al. (1977) have measured the ellipsoid axes of the crown envelope for maize and found the ratio of these axes $h/2 : R_x : R_y$ to be 3.8:2.4:1 (Fig. 1). At the same time for horse beans the ratio was 3.5:1.5:1.

Having determined the crown envelope, we must derive the following characteristics: (1) the projection area of the crown envelope $S(z, \Omega)$ on a horizontal plane at the height z in the direction $\Omega = (\theta, \phi)$; (2) the volume of the crown envelope

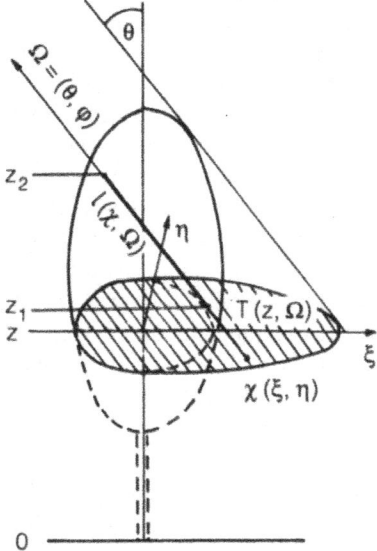

Fig. 2. The projection region $T(z, \Omega)$ of the crown envelope on a horizontal plane at the height z in the direction $\Omega = (\theta, \phi)$ and the ray path length $l(X, \Omega)$ inside the crown envelope

V(z) above the height z; (3) the entry and exit points of a ray for each required inclination, azimuth and position and the respective path length of the ray in the crown envelope (Fig. 2).

Formulas for these characteristics are rather lengthy even if simple geometrical figures are involved (e.g. Nilson 1977; Norman and Welles 1983). However, modern computers do not need analytic expressions to calculate the area of the shadow cast by given figures.

3.2 Foliage Area Distribution in the Crown Envelope

The distribution of the foliage area within the crown envelope is essential in considering radiation penetration and interception. Methods to determine foliage area distribution have been reviewed by Ross (1981) and Myneni et al. (1989).

Here, the 3-D foliage area distribution is considered. Following Ross (1981), define the foliage area volume density distribution $u(X)$ (m^{-1}) as the amount of (one-sided) foliage area at the point $X(x, y, z)$ per unit volume area. Provided a certain plant organ needs to be particularly pointed out, a respective subscript is used, e.g. u_L for leaves. The integral

$$s = \iiint_V u(X) dx\, dy\, dz$$

will yield the total foliage area (m^2) of the plant, V being the spatial region of the crown envelope.

An equivalent representation of the 3-D distribution of the foliage area is obtained provided the respective probability distribution function, which is normalized to unity together with the total foliage area s, is given (e.g. Mann et al.

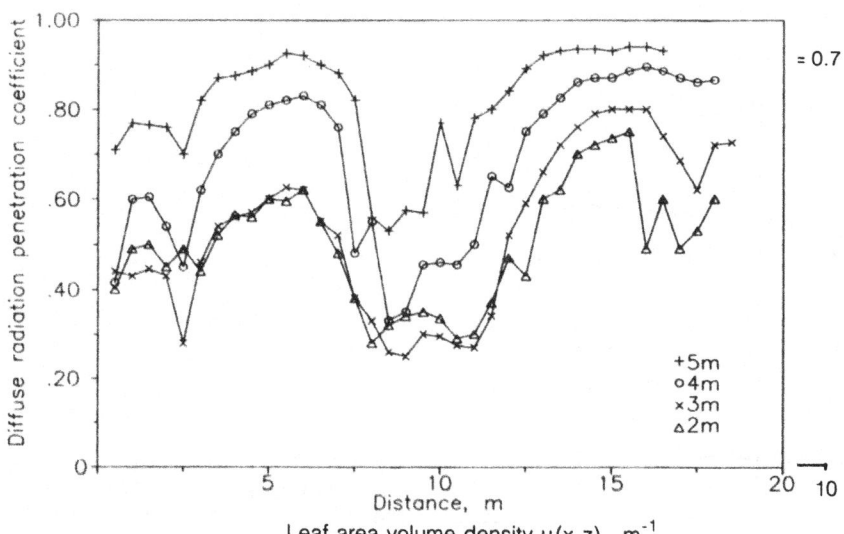

Fig. 3. Two-dimensional distribution of the leaf area volume density $u(x, z)$ in maize, determined as an average over 11 plants using the point quadrat method (Nilson et al. 1977)

1979). Dividing the total plant foliage area s by the projection area of the crown envelope in the vertical direction, $S(0,0)$, the foliage area index of the plant is obtained.

In most of the light penetration models the foliage area is assumed to be uniformly distributed ($u = \text{const}$) within the crown envelope (Charles-Edwards and Thornley 1973; Oker-Blom and Kellomäki 1982; Nilson 1984; etc.). Usually, these models seem to work reasonably well. However, the investigators who have measured the spatial distribution of u have found it, as a rule, rather far from uniform. Some of the measurement results for saltbush (Philip 1965; Warren Wilson 1965) and maize (Daynard 1971; Nilson et al. 1977; see Fig. 3) have indicated that the leaf area density decreases towards the crown periphery. However, for many trees the highest leaf area densities are at the crown periphery, while the crown interiors may be devoid of leaves (but not without the woody part). This is just the case when it seems reasonable to divide the crown envelope into subcanopies with different uniform foliage area densities. This has been proposed by Norman and Welles (1983) in their general array model (GAR) and by Koppel and Oja (1984) in a Norway spruce crown model.

Detailed information on the distribution of foliage weight and area in some tree species has been given by Dylis and Nosova (1977) and Brown (1978). Kuuluvainen et al. (1988) studied the clustering of shoots into whorls in 8-year-old Scots pine trees. They noted that some of the plants studied showed a regular undulation of the vertical distribution of needle mass. The undulation appeared to be related to the whorls. Between individual whorls there were spaces devoid of needles. An adequate foliage area distribution function must consider such a phenomenon, too. Therefore, a correct model for the vertical distribution of the needle area in those trees should follow curve (1) in Fig. 4 rather than the smooth curve (2). The effect of a regularly undulating vertical distribution of foliage on light penetration is easy to imagine, particularly at low sun elevations.

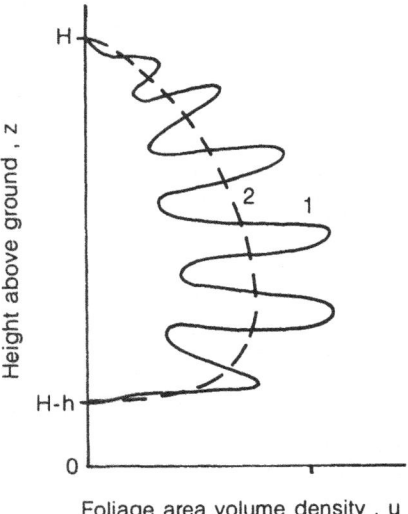

Fig. 4. Vertical distribution of the needle area volume density in a pine tree. A hand-drawn figure inspired by the results of Kuuluvaninen et al. (1988). Curve *1* The clustering of shoots into whorls; *2* the smoothed curve

3.3 Inclination and Azimuth Distribution of Foliage Elements

The treatment of foliage inclination and azimuthal distribution, as proposed for homogeneous crops is also applicable to individual plants. So, define $g(X, \Omega_L)d\Omega_L/(2\pi)$ as the probability to observe the normal of a foliage element to be directed into the solid angle $d\Omega_L$ around the direction $\Omega_L = (\theta_L, \phi_L)$, θ_L being the normal's zenith angle. The distribution function g may vary within the crown envelope. Height dependence of the function g has been reported by various authors, but there is practically no experimental evidence available on the 3-D variability. Evidently, the maize in Fig. 1 shows at least 2-D variability.

Nonuniform leaf azimuth distributions are common in isolated plants. For instance, in maize plants having a phyllotaxis with the genetic spiral angle $\alpha = 180°$, the leaves tend to be oriented nearly in one plane. Nonuniform azimuth distributions have also been reported for some plants grown in rows, while distributions with clear daily courses have been recorded in heliotropic plants.

There is a great (but insufficient) body of literature on leaf inclination distributions. Several theoretical distributions have been proposed to simulate those measured: spherical (Nichiporovich 1961), planophile, erectophile, plagiophile (de Wit 1965), beta-distribution (Goel and Strebel 1984), elliptical distribution (Campbell 1986), etc. Von Mises circular distribution was used by Shell and Lang (1975) to describe leaf azimuth and inclination distributions and their heliotropism-related changes in the course of the day.

Two important geometrical and optical canopy characteristics are closely related to the foliage orientation. As defined by Ross (1981), these characteristics are: the mean projection of the unit foliage area onto a plane perpendicular to the direction Ω

$$G(\Omega) = \int_0^{2\pi} \int_0^{\pi/2} g(\Omega_L)|\Omega \cdot \Omega_L|\sin\theta_L \, d\theta_L \, d\phi_L/(2\pi), \tag{1}$$

and the (area) scattering phase function of the canopy medium for the incident ray direction Ω' and scattered ray direction Ω

$$\Gamma(\Omega' \to \Omega) = \int_0^{2\pi} \int_0^{\pi/2} g(\Omega_L)\gamma(\Omega_L, \Omega' \to \Omega)|\Omega \cdot \Omega_L||\Omega' \cdot \Omega_L|\sin\theta_L \, d\theta_L \, d\phi_L/(2\pi), \tag{2}$$

where $\gamma(\Omega_L, \Omega' \to \Omega)$ is the scattering phase function for an individual foliage element; $\Omega \cdot \Omega_L = \cos\theta\cos\theta_L + \sin\theta\sin\theta_L\cos(\phi - \phi_L)$. For the phase function γ, the so-called bi-Lambertian approximation is the most widely used, assuming γ to be equal to a constant, r, for reflection and to another constant, t, for transmission. However, in recent works, in addition to Lambertian reflection, a specular component is considered (Vanderbilt and Grant 1985; Nilson and Kuusk 1989). Definitely, both the characteristics G and Γ may be functions of the location inside the canopy envelope if the distribution function g and/or the scattering phase function γ depend on location.

Knowing the function G, the total projected foliage area of the plant onto a plane perpendicular to the direction Ω will be

$$S_1^*(\Omega) = \iiint_V u(x, y, z)G(x, y, z, \Omega) \, dx \, dy \, dz, \tag{3}$$

where V denotes the spatial region of the canopy envelope. The projection onto a horizontal plane is

$$S^*(\Omega) = S^*_\perp(\Omega)/\cos\theta. \tag{4}$$

$S^*_\perp$ and S^* may be defined also as functions of the height z, then the respective crown envelope is chopped from below at the height z. The ratio of the total projected foliage area $S^*(z, \Omega)$ to the respective projection area of the crown envelope $S(z, \Omega)$ characterizes the mean number of interceptions at the height z in the view direction Ω.

3.4 Proportion of Gaps

It is convenient to define the binary random function characterizing the presence of gaps in the foliage as

$$\alpha_1(X, \Omega) = \begin{cases} 1, & \text{for a free line-of-sight from the point} \\ & X(x, y, z) \text{ in the direction } \Omega = (\theta, \phi); \\ 0, & \text{if the line-of-sight is intercepted by any of the} \\ & \text{plant foliage elements.} \end{cases} \tag{5}$$

The point X may be either inside or outside the crown envelope. The mean value of this binary random function, $\bar{\alpha}_1(X, \Omega)$, is equal to the average proportion of gaps in the crown as viewed from the point X in the direction Ω. Consequently, it serves as a coefficient of radiation penetration.

Mann et al. (1979) have proposed a general law of direct sunlight penetration. Starting with the ideas presented in Mann et al. (1977), they have considered both the variable path length within the crown envelope and the nonuniform distribution of foliage within the crown envelope. The average proportion of gaps in the crown envelope projection region $T(z, \Omega)$ (Fig. 2) at the height z in the direction Ω is given as

$$a_1(z, \Omega) = \iint\limits_{T(z, \Omega)} \exp[-c_\Omega f_\Omega(\xi, \eta)] d\xi d\eta/S(z, \Omega), \tag{6}$$

$S(z, \Omega)$ being the area of the projection region $T(z, \Omega)$, c_Ω is the total projected foliage area, $f_\Omega(\xi, \eta)$ is the probability distribution density of foliage elements projected onto the horizontal plane at the height z, ξ and η are the horizontal coordinates associated with the projection region T. Here, the basic idea lies in calculating the gap probability separately for the each point (ξ, η) in the projection region T and later averaging the gap probability over the region T.

Knowing the volume density distribution of the foliage area u and the foliage inclination and azimuth distribution g, calculation of $c_\Omega \cdot f_\Omega$ could be reduced to integration of the product $u \cdot G$ along the path of the ray within the canopy envelope, while G is given by formula (1). Following the present notation, formula (6) may be written as

$$a_1(z, \Omega) = \iint\limits_{T(z, \Omega)} \bar{\alpha}_1(\xi, \eta, z, \Omega) d\xi d\eta/S(z, \Omega) \tag{7}$$

and

$$\bar{\alpha}_1(\xi,\eta,z,\Omega) = \exp\left[-\int_0^1 u(x,y,z)G(x,y,z,\Omega)d1 \right], \tag{8}$$

where $1 = 1(\xi,\eta,z,\Omega)$ stands for the respective path length (see Fig. 2). The use of simple geometrical relations yields instead of Eq. (8) the following equation

$$\bar{\alpha}_1 = \exp\left\{ -\int_{z_1}^{z_2} u[x + (z' - z)\tan\theta\cos\phi, y + (z' - z)\tan\theta\sin\phi, z']G\,dz' \right\}, \tag{9}$$

where z_1 and z_2 correspond to the entry and exit points of the ray. For rotationally symmetrical crown envelopes one may take $\phi = 0$. Formula (8) and the homogeneous layer analogue have been used by Ross and his colleagues since the mid-1960s (e.g. Ross and Nilson 1965).

If the volume density of the foliage area u and the orientation function g are supposed to be constants throughout the crown envelope, then formula (8) is reduced to

$$\bar{\alpha}_1(\xi,\eta,z,\Omega) = \exp[-\tau(\Omega)1(\xi,\eta,z,\Omega)], \tag{10}$$

where 1 denotes the path length inside the crown envelope of the line-of-sight, initiating from the point (ξ,η,z) in the direction Ω, and $\tau(\Omega) = uG(\Omega)$ is the extinction coefficient per unit path length.

The mean path length, if averaged over the whole projection region T of the crown envelope, may be calculated as

$$\bar{1}(z,\Omega) = V(z)/[S(z,\Omega)\cos\theta]. \tag{11}$$

The mean penetration coefficient averaged over the region T may be given approximately as

$$a_1(z,\Omega) = \exp[-\tau(\Omega)\bar{1}(z,\Omega)], \tag{12}$$

which underestimates the penetration coefficient since the variable path length is ignored (see Mann et al. 1979). Treating the function $\alpha_1(\xi,\eta)$ in formula (10) as a function of the random argument 1, then the procedure of approximate statistical linearization may be applied [formula (27), below]. Thus, we obtain an approximation

$$a_1(z,\Omega) \approx \exp[-\tau(\Omega)\bar{1}(z,\Omega)][1 + \tau^2(\Omega)\sigma_1^2(z,\Omega)/2], \tag{13}$$

where $\sigma_1^2(z,\Omega)$ is the variance of the path length over the region T.

3.4.1 Spatial Dispersion of Foliage

A straightforward interpretation of formula (8) indicates that it gives us the probability of zero interceptions by the foliage elements over the whole path length 1 with the ray in the direction Ω. The expression in the exponent $\int uGdl$ represents the mean number of interceptions. The main assumption in formula (8) is that interceptions of the ray occur entirely independently of each other, according to the spatial Poisson distribution. Note that the Poisson parameter – the mean number of interceptions per unit path length – is allowed to vary spatially.

The basic assumption in the Poisson model – random and independent dispersion of foliage may seem rather doubtful. Defining u(X) as the mean volume density distribution of the foliage area, no information about the spatial correlation of foliage location has been given. The same also holds for the foliage orientation. A correlation function of the foliage element locations along the ray path is needed. At least two kinds of correlations are easy to imagine: positive for the presence of clustered and sparser regions, and negative for the foliage elements with the tendency to avoid overlapping. Thus, the coefficient of correlation is a function of the distance between the two points under study.

Unfortunately, almost no experimental data on these correlations are available. Theoretical methods extending the Poisson distribution-based treatment use other probability distributions to account for regularity or clumping (e.g. Acock et al. 1970; Nilson 1971). The simplest way, however, seems to introduce a correction coefficient into the exponent in formula (8), whereas the rest of the basic formula remains unchanged. This correction coefficient is less than unity for clumping and more than unity for the regular dispersion of foliage. This means simply that the density of the foliage area has to be decreased or increased to account for clumping or regularity, respectively. In formula (6), proposed by Mann et al. (1979), the clumping or regularity effects should already be included in the distribution of projected foliage f_Ω.

There is no definite answer to the question about the role of structural regularities of the foliage in radiation penetration. The plant-leaf arrangement as a certain light-catching system is likely to contain considerable regularities. But daily solar movement makes it nearly impossible to optimize the plant structure in a simple way. Some plants try to optimize light interception mostly by heliotropic leaf angle movements and less by changing their locations. Only under fairly permanent illumination conditions, can the proper leaf positioning yield a considerable effect on productivity. In fact, quite often we can see complete leaf mosaics among those plants.

Plants frequently show structural regularities in the vertical view direction. On the one hand, this may be explained by the fact that plants grow chiefly in the vertical direction. On the other hand, the maximum diffuse radiation is received from the near-zenith directions, which are more open to radiation penetration in the community. Light interception in these directions depends markedly on the plant structure, while inclined directions are often obscured by adjacent plants. Note that foliage structures, being regular in the vertical view direction, may be treated as random for those in the inclined direction. Such an effect has been confirmed by Monte-Carlo simulations (Ross and Marshak 1987) in canopy reflectance calculations.

In contrast, clumping (clustering, grouping) of foliage is a very common phenomenon in plants, particulary in conifer trees. Norman and Jarvis (1975) have shown that it was just the clustering of needles into shoots in Sitka spruce that caused the major part of the clustering effect on radiation penetration. Clustering of branches into whorls and the spatial pattern of shoot distribution in the crown envelope were of less importance.

There are two relatively simple ways to consider the influence of clustering on gap frequency. If clusters are located regularly, such as whorls is some coniferous trees (Fig. 4), then the 3-D distribution of the foliage area density may be defined in accordance with these clusters. Using gap frequency formulas (6–9), this type of clustering is included automatically. Alternatively, a proper angle-dependent correction term (< 1) may be introduced into the exponent in formula (10). If the clusters are considered as those randomly dispersed, then only the second method may be applied. Oker-Blom and Smolander (1988) have shown that grouping of needles into shoots in Scots pine reduced the radiation extinction coefficient by a factor 0.57.

Introducing a correction term for clustering of needles into shoots is equivalent to considering shoots as the smallest foliage elements in conifers. So, instead of the volume density of the needle area, the volume density of the shoot number together with the respective shoot structure model is considered. Here, it is namely the concept of smaller subcanopies that seems to be useful. A model of shoot structure should contain exactly the same components as that of plant structure: shoot envelope, the volume density distribution of needle area within the shoot envelope, needle inclination and azimuth distribution and perhaps an index of the spatial distribution pattern for the needles.

3.4.2 Bidirectional Gap Probability

Up to now, no information on the linear dimensions of the foliage elements has been used in gap frequency modeling. Now assume that we have been able to calculate gap proportions for any point and view direction of interest. We need bidirectional gap probabilities, too, Define as bidirectional gap probability, $Q_1(X, \Omega, \Omega')$, the probability that when viewed simultaneously from the point X in two directions, Ω and Ω', both the respective lines-of-sight are not obscured by any foliage elements. In fact, bidirectional gap probabilities deserve much more attention. Only recently, in connection with the theoretical consideration of the hot-spot phenomenon in the directional distribution of canopy reflectance, bidirectional gap probabilities became actual.

Logically, if the directions Ω and Ω' are sufficiently apart from each other, gap probabilities in these directions may be treated as independent random variables, i.e. $Q_1(X, \Omega, \Omega') = \bar{a}_1(X, \Omega) \cdot \bar{a}_1(X, \Omega')$. Provided directions Ω and Ω' are close to each other, then gap probabilities are at least partly mutually dependent. For the exact backward direction $Q_1(X, \Omega, \Omega) = \bar{a}_1(X, \Omega)$.

A correction factor, C_{HS}, is introduced for the correlation between the gap frequencies in the two directions involved (e.g. Nilson and Kuusk 1985):

$$Q_1(X, \Omega, \Omega') = \bar{a}_1(X, \Omega)\bar{a}_1(X, \Omega')C_{HS}(X, \Omega, \Omega'). \tag{14}$$

The main reason for the nonzero correlation lies in the finite angular size of foliage elements and gaps; the same element may simultaneously cause shading in the two directions Ω and Ω' and both the lines-of-sight may occur in the same gap. Kuusk (1985) has derived theoretical formulas of the function C_{HS} for a layered canopy. In addition to the traditional canopy structural parameters (u, G), a typical foliage element size is introduced. The essence of this treatment lies in considering a correlation function of the foliage element location in the horizontal direction. Although any kind of correlation function can be used, for practical purposes, a simple exponential correlation function $r(\delta) = \exp(-\delta/w)$ is introduced. In this expression δ stands for the correlation lag (horizontal distance between the two points of interest) and w for a correlation radius, characterizing the speed of correlation attenuation. Provided statistically independent locations of foliage elements are assumed in a horizontal layer, w is determined by the element and gap size only. When two rays of different directions are considered simultaneously, the rays travelling upwards from the same point through the canopy form gradually increasing correlation lag values. Consequently, the nearest layers contribute most to the correlation. Besides the element size, the degree of statistical dependence of gap frequencies in two directions and the factor C_{HS} in formula (14) are determined by the angle β between these two directions.

According to the C_{HS} determination in formula (14), $C_{HS} = 1/\bar{\alpha}_1$, if $\beta = 0$. As a rule, there is a rapid decrease in C_{HS} at small β values, while for greater values the decrease is reduced. Our understanding of the angular correlation functions of gap frequency in various plants is rather limited. From the above general considerations it may be concluded that a typical element size is essential in forming the shape of the correlation function. Note that the magnitude of correlation is not merely determined by the size of the smallest element. So, for conifer trees a rapid decrease is likely in the correlation function already at very small angles between two directions; it can be associated with the effect of needle size. Dimensions of other structural units, such as shoots, branches, and trunk, are also likely to contribute to the correlation behaviour.

3.5 Penetration of Direct Solar and Diffuse Sky Radiation

3.5.1 Direct Solar Radiation

If the sun were a point source, then direct solar radiation might be treated as a parallel beam of radiation. Then the respective penetration coefficient as a random variable would be only two-valued, 0 or 1, and just equal to the random gap frequency variable $\alpha_1(X, \Omega)$ as defined by formula (5).

Due to the finite angular size of the solar disc the penetration coefficient of direct solar radiation $a_s(X, \Omega_0)$ can be expressed as the following integral:

$$a_s(X, \Omega_0) = \iint\limits_{T_s(\Omega_0)} B_s(\Omega)\alpha_1(X, \Omega) \cos\theta \sin\theta \, d\theta \, d\phi, \qquad (15)$$

where $T_s(\Omega_0)$ denotes the angular region corresponding to the apparent solar disc, $B_s(\Omega)$ is the normalized radiance distribution over the solar disc, so that

$$\iint\limits_{T_s(\Omega_0)} B_s(\Omega) \cos\theta \sin\theta \, d\theta \, d\phi = 1.$$

Mostly, $B_s(\Omega)$ is set to a constant.

In spite of the small angular dimensions of the solar disc, in many practical cases the value of the random function $\alpha_1(X, \Omega)$ cannot be considered a constant,

0 or 1, throughout the region T_s. If only part of the solar disc is obscured, the phenomenon of penumbra occurs. Although a quite simple phenomenon by nature, a quantitative description of penumbra is rather complicated (e.g. Norman et al. 1971; Denholm 1981).

Calculation of the mean coefficient of direct solar radiation penetration in formula (15) requires the replacement of a_s and α_1 by the respective expectation values. Since angular variations in $\bar{\alpha}_1$ are of rather low frequency, it may be set to $\bar{a}_s = \bar{\alpha}_1$. But in calculating the distribution function of the penetration coefficient a_s by formula (15), the role of the angular correlation of the gap frequency becomes evident. Since very small angular differences ($< 0.5°$) are involved in Eq. (15), we have to consider rather fine angular structures with regards to the correlation.

We simulate the penumbra problem in a simplified manner. Although over-simplified, this approach gives us some idea of the qualitative changes in the distribution of direct sunlight penetration both in single plants and in plant canopies. Assume that instead of studying thoroughly the presence of gaps in all the directions corresponding to the solar disc, we choose N discrete directions within the region of the solar disc. Further, assume that each of these directions is responsible for an equal part of the disc and in each of the directions only a binary-valued gap probability is possible. If the gap probabilities in all chosen directions could be treated as independent random variables, then it can easily be seen that the direct sunlight penetration is described by the binomial distribution:

$$P(i/N) = \frac{N!}{i!(N-i)!}(1 - \bar{a}_s)^{N-i}(\bar{a}_s)^i, \quad i = 0, 1, \ldots, N, \tag{16}$$

where $P(i/N)$ denotes that the probability of the direct solar radiation penetration coefficient is equal to i/N, $\bar{a}_s$ being the mean value of the penetration coefficient.

For a plant with large leaves the angular correlation function decreases slowly. Thus, the apparent solar disc cannot be subdivided into independent subregions. Hence, $N = 1$ and only two values of the penetration coefficient are possible and we have no penumbra: $P(0) = 1 - \bar{a}_s$, $P(1) = \bar{a}_s$.

In contrast, for a plant with very small, randomly dispersed foliage elements located at nearly the same level and having no clusters, the angular correlation function drops rapidly to zero and the number of independent directions may be quite large. Formula (16) gives an approximation of the normal distribution function. Of course, intermediate cases are possible with moderate N values.

Unfortunately, this simulation does not work well in many real situations, e.g. with clusters and foliage elements having considerable vertical dispersion and with mixtures of elements of different sizes. Then the angular correlation function drops rather rapidly at small values of the angular lag. At the further increase in the angular lag the decrease in correlation may slow down. Therefore, it is impossible to choose a proper subdivision of the apparent solar disc with entirely independent discrete directions. Partial dependence with a positive correlation, which is quite common, gives rise to an increase in the variability of the direct solar radiation penetration coefficient. As a result, in practice, the distributions measured differ from those modeled by the trivial binomial distribution mostly in larger frequencies near the extreme classes (full sun, full shade) of the penetration coefficient.

More correct in a mathematical sense would be to treat the distribution of the direct solar radiation penetration coefficient as the sum of many partly correlated binary random variables. For more details on the theoretical considerations related to the penumbra, the reader is referred to papers by Denholm (1981), Anisimov and Menzhulin (1983), Oker-Blom (1984) and Myneni and Impens (1985).

3.5.2 Diffuse Sky Radiation

The penetration of diffuse sky radiation into the canopy of a plant can be treated similarly to that of direct solar radiation. The only difference is that the radiation source is hemispherical. A random value of the diffuse radiation penetration coefficient is calculated as

$$a_D(X) = \int_0^{2\pi} \int_0^{\pi/2} I_D(\Omega)\alpha_1(X,\Omega)\sin\theta\cos\theta\,d\theta\,d\phi, \tag{17}$$

where $I_D(\Omega)$ denotes the normalized distribution of the sky radiance. Here, $\alpha_1(X,\Omega)$ is the random distribution of gaps in the foliage over the whole hemisphere. This distribution is possible to imagine as a contrasting black-and-white hemispheric photo taken at the point X. To obtain the mean value of a_D, the angular gap distribution α_1 must be replaced by its expectation $\bar{\alpha}_1$. For theoretical estimates of $\bar{\alpha}_1$, again formulas (8) or (9) can be used.

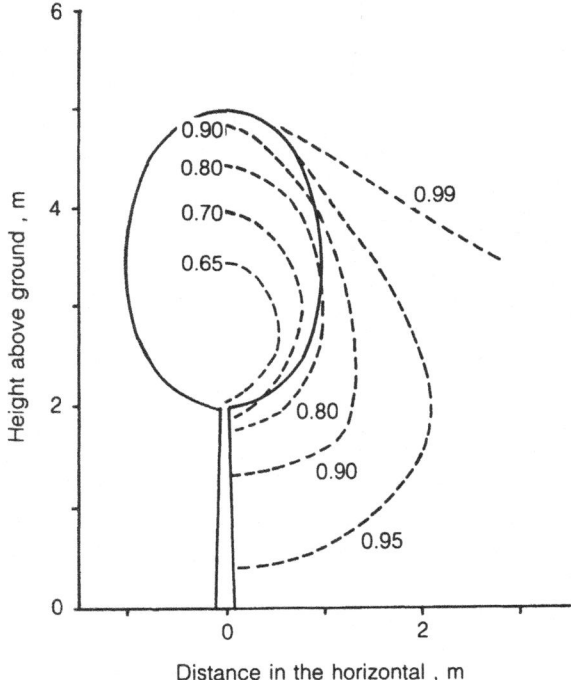

Fig. 5. Isolines of the diffuse sky radiation penetration coefficient a_D formed by an ellipsoidal crown of a single tree

Figure 5 is a schematic representation of the calculated isolines of the diffuse radiation penetration coefficients a_D for a single tree crown. The tree crown envelope was modeled as an ellipsoid of rotation, foliage area was assumed to be uniformly distributed in the crown envelope ($u = 0.72\,\mathrm{m}^{-1}$), and the foliage angle distribution was assumed to be spherical. Surprisingly, the effect of a single tree crown on the diffuse radiation penetration diminishes rapidly with increasing distance from the tree. Some aspects of single tree-induced, diffuse, sky radiation penetration are discussed in Koppel and Oja (1984).

To obtain information on the distribution function of the diffuse skylight penetration coefficient, the correlation function of the gap frequency is required. For the variance of a_D we have

$$\sigma^2(X) = \int_0^{2\pi} \int_0^{\pi/2} \int_0^{2\pi} \int_0^{\pi/2} I_D(\Omega) I_D(\Omega') K_\alpha(X, \Omega, \Omega') \sin\theta \cos\theta \sin\theta' \cos\theta'\, d\theta\, d\phi\, d\theta'\, d\phi',$$

where $K_\alpha(X, \Omega, \Omega')$ is the covariation function of the gap frequency at the point X in the directions Ω and Ω'. Here, we need both the fine angular structure of the correlation function and also the rougher structure.

3.6 Modeling of Radiation Scattering

Most of the models for the radiation regime in the nonhomogeneous plant canopies ignore the portion of solar radiation scattered by foliage elements. The number of papers on radiation scattering within isolated plants is small. The problems of radiation scattering within limited canopy envelopes seem to have little relevance, since the contribution of the scattered radiation in the total PAR is small. This point of view may be accepted in photosynthetic studies, but not in remote sensing or albedo calculations.

The common theoretical basis for describing radiation scattering within a plant medium is the radiative transfer equation. This equation, originally derived for the transfer of radiation is stellar and planetary atmospheres (Chandrasekhar 1960) and for neutron transport (Davison 1958), has been modified by Ross (1964, 1981) for use in plant canopies. In this integro-differential transfer equation the function G [formula (1)] appears as a direction-dependent extinction coefficient, while the function Γ [formula (2)] is a scattering phase function for the canopy medium. Extensive reviews on the problems concerning the radiative transfer equation have been given in Ross (1981), Myneni et al. (1989) and Myneni and Ross (1991). Many authors who have modeled radiation scattering in plant canopies do not refer to the transfer equation but in fact use some version of its approximations.

Note that in using the transfer equation in a single plant, apart from presenting the same problem in a homogeneous canopy infinitely extended in the horizontal, the problem is essentially three-dimensional. In other words, provided that the distribution of the foliage area is assumed to be uniform over the whole single plant crown envelope, we have to solve a 3-D transfer equation. Because of the finite horizontal extension of the canopy, there always exist horizontal gradients of the radiation field, too. The three-dimensionality makes computation rather complex, in particular, when multiple scattering is concerned. Still more complications arise from boundary conditions, e.g. those caused by multiple interactions between the plant and the underlying ground.

The Monte Carlo method, when applied to canopies of finite horizontal extension, seems to be one of the most effective approaches to the problem (e.g. Kimes and Kirchner 1982). Kanevski and Ross (1985) used the Monte Carlo method to simulate the angular distribution of reflectance caused by a single spruce tree. They found the role of multiple scattering to be rather significant in the near-infrared region of the spectrum, i.e. up to 50% of the total reflectance. According to Kanevski and Ross (1985), to achieve an accuracy of 95% of the reflectance, the scattering orders up to the third (incl.) should be considered in the near-infrared region.

The relatively small number of scattering orders needed encourages one to use the method of successive-order iterations (see Myneni et al. 1989) to solve the 3-D transfer equation for individual plants. Recently, Myneni (pers. commun.) used a discrete ordinate method to solve the 3-D transfer equation in canopies of finite horizontal extension.

Kuusk (1987) proposed a model for calculating the angular distribution of radiation reflected by a single tree canopy. The model is based on first-order scattering calculations. The tree crown envelope, modeled as an ellipsoid of rotation, is divided into small blocks. For each block, the bidirectional gap probability Q_1 [formula (14)] is calculated. The reflectance factor, $\rho(\Omega_0, \Omega)$, of the tree crown averaged over the whole projection area of the crown envelope $S(0, \Omega)$ in the direction Ω with the incident direct solar radiation direction being Ω_0 may be defined. It can be expressed as

$$\rho(\Omega_0, \Omega) = \iiint_V \Gamma(X, \Omega_0 \to \Omega) Q_1(X, \Omega_0, \Omega) u(X) \, dx \, dy \, dz / [S(0, \Omega) \cos \theta_0], \qquad (18)$$

where Γ denotes the scattering phase function of the canopy medium and V is the spatial region corresponding to the crown envelope. Kuusk's calculations showed that besides the expected effect of the projection areas of the crown envelope as viewed from the directions Ω and Ω_0, the angular dependence of the reflectance was greatly influenced by the bidirectional gap probability Q_1. Introduction of Q_1 made it possible to simulate a sharp maximum in the backscattering direction, the hot spot effect. Kuusk's approach appears to be also directly applicable to downward radiation scattering.

It is of interest to note that it is essential to consider the problem of radiation scattering for the structural elements below the plant level also. In particular, for remote sensing purposes a scattering phase function for coniferous shoots is needed. This phase function model of a shoot may then be used as a submodel to create respective models for a single crown or for the whole forest canopy reflectance.

4 Plant Stand. Modeling of Stand Structure and Radiation Regime

For a plant stand, coordinates x, y, z, which are associated with the stand, are introduced. For row structures, the x-axis is supposed to be directed across the rows and the y-axis is assumed to proceed along the rows, with the origin of the coordinates located at the ground in the centre of a row.

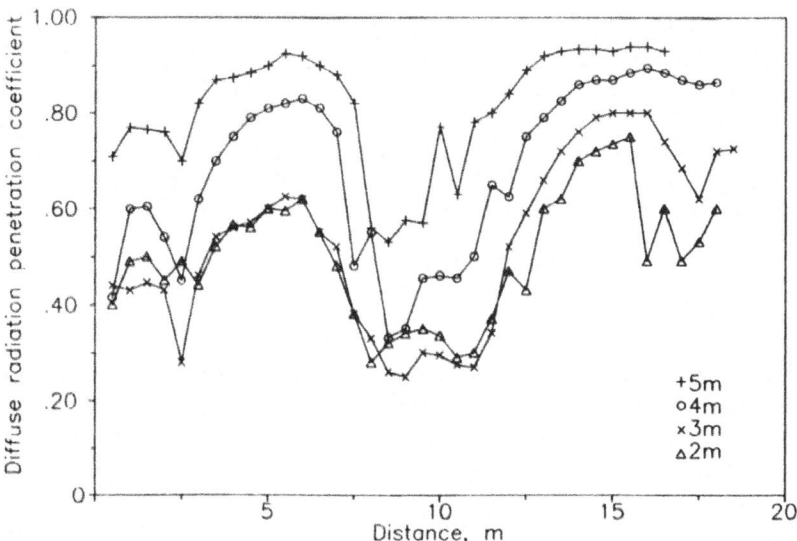

Fig. 6. Horizontal profiles of diffuse radiation penetration coefficients in a 30-year-old, naturally regenerated Scots pine stand (Calluna type). Estimations from hemispherical photographs at a height of 2, 3, 4 and 5 m (Hari et al. 1985)

There are no principal differences in calculating the penetration of direct solar and diffuse sky radiation for plant stands compared with a single plant. Formulas (15) and (17) hold for plant canopies as well. The key problem is the evaluation of gap frequency. Figure 6 shows an example of the horizontal variability of the diffuse radiation penetration coefficient in a heterogeneous Scots pine forest.

4.1 Models of Gap Proportion

4.1.1 Computer Models

According to the above concept, a plant canopy is modeled as an array of individual subcanopies, i.e. plants. In order to completely define the plant canopy, the parameters of all the plants forming the canopy as well as the arrangement pattern of these plants are required. Such models can easily be applied to some forest plots, orchards or agricultural crops where each plant has been measured and plant locations mapped. Then it is possible to calculate gap probabilities for each point in the canopy and for any view direction. The calculation procedure is straightforward but rather time-consuming (e.g. Norman and Welles 1983; Nilson 1984; Hari et al. 1985; Welles and Norman 1991).

For each point $X(x, y, z)$ under study within the plant canopy a set of lines-of-sight (view directions) is chosen to characterize the daily sun movement and/or the whole upper hemisphere for sky radiation. For each line-of-sight, all the plants with crown envelopes intersected by the line-of-sight and the respective entry and exit points are determined. Making use of the gap probability formulas (6), (7) or (12), the coefficients of penetration are determined separately for each plant intersected. Total gap probability $a(X, \Omega)$ at the point X in the direction Ω will be obtained as the product of gap probabilities in the individual crown envelopes involved. This means that the locations of foliage elements in different crown envelopes are supposed to be independent. Since the expressions for the gap probabilities are exponential functions, the resulting product is also exponential. For instance, we have

$$a(X, \Omega) = \exp\left[-\sum_i l_i(X, \Omega)\tau_i(\Omega) \right], \tag{19}$$

if the volume density of the foliage area u and orientation G are assumed to be constants within each crown envelope, $l_i(X, \Omega)$ being the path length of the line-of-sight within the i-th crown envelope, $\tau_i(\Omega) = u_i G_i(\Omega)$ is the radiation extinction coefficient per unit length; summation is carried out over all the intersections. Here, one must foresee the situation when the crown envelopes of adjacent plants overlap. The problem is how to define the extinction coefficient in the region of overlapping. In principle, three different overlapping situations may occur:

1. the foliage area volume density in overlapped regions is the sum of the respective volume densities of individual plants;
2. volume density in overlapped regions is zero (e.g. as the result of a mechanical interaction);
3. an intermediate case with partial interaction.

In the same way, canopies of regularly spaced plants such as row or square-sown crops may be modeled (Norman and Welles 1983; Whitfield 1986). For row crops with the plants more or less randomly located in rows, the canopy may be considered as homogeneous along the y-axis (row direction) but inhomogeneous along the x-axis (across the rows). This inhomogeneity might be described either by defining subcanopies corresponding to the rows (e.g. Charles-Edwards and Thorpe 1976) or by defining a 2-D and periodic foliage area distribution u(x, z) (Allen 1974; Fukai and Loomis 1976; Nilson et al. 1977). Assuming $G = G(z, \Omega)$ in the latter case, the proportion of gaps $a(x, z, \theta, \phi)$ at the point (x, z) in the direction $\Omega = (\theta, \phi)$ is

$$a(x, z, \theta, \phi) = \exp\left\{ -\int_z^H u[x + (z' - z)\tan\theta\sin\phi, z']G(z', \theta, \phi)dz'/\cos\theta \right\}, \tag{20}$$

where H is the height of the canopy; $\phi = 0$ corresponds to the row direction. Note that u(x, z) may be determined when the single plant crown envelope and the foliage area distribution within it, together with some plant location parameters, are known.

4.1.2 Analytical Model

Nilson (1977) proposed an analytical model, sometimes called the "silhouette model", to simulate the gap probability and radiation penetration into plant canopies. A very similar approach has been reported by Kuuluvainen and Pukkala (1987). They introduced a "shadow surface" to illustrate the local shading effect in the forest; the probability of shading caused either by a single tree or by the whole canopy. The shadow surfaces were calculated for momentary shading and as averages for longer periods (day, growing season). The influence of the crown height-radius ratios and the tree distribution pattern (Poisson, regular) on the shadow surfaces was studied. Starting with the same ideas, Li and Strahler (1988) suggested a model for the gap probability of a discontinuous vegetation canopy.

Now consider Nilson's model in greater detail. First, consider a plant stand consisting of identical plants. Suppose our aim is to calculate the average proportion of gaps $a(z, \Omega)$ in the canopy at the height z in the view direction Ω (Fig. 7). When viewed from a point at the height z the respective line-of-sight may be non-obscured in the following, mutually excluding cases:

1. The line-of-sight does not intersect any of the crown envelopes (or stems), i.e. we have a gap between plants. In such a case no plant stems should occur within a certain "vulnerability" region (hatched area in Fig. 7). The area of this region is equal to the crown envelope (+ stem) projection area $S(z, \Omega)$ onto a horizontal plane at the height z and projected by the rays at the zenith angle θ and azimuth angle $\phi' = \phi + \pi$.

2. The line-of-sight is intersected by one crown envelope only, i.e. we have a gap inside a crown. In this case one and only one plant stem is located in the vulnerability region.

3. The line-of-sight is intersected by two plant crowns, i.e. we have a gap in the array of two crown envelopes. Two plant stems should occur in the vulnerability region, etc.

Such an approach yields the following expression for the gap proportion:

$$a(z, \Omega) = P_0(S) + a_1 P_1(S) + a_2 P_2(S) + \cdots, \tag{21}$$

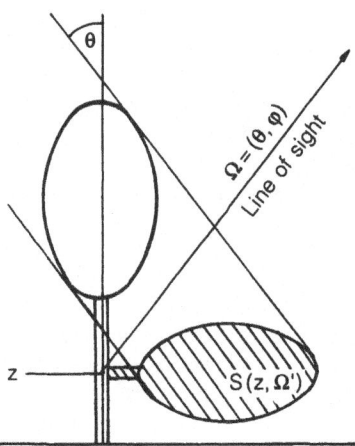

Fig. 7. Derivation of the analytical expression for the gap proportion in a plant canopy

where $P_i(S)$ denotes the probability of i plant stems occurring within the region of area $S = S(z, \Omega)$, equal to the projection area of the crown envelope, $a_1 = a_1(z, \Omega)$ is the mean proportion of gaps in a single plant crown, as viewed from the height z in the direction Ω [formulas (6), (7) or (12)], $a_2 = a_2(z, \Omega)$ is the average proportion of gaps for an array of two plants, etc.

Note that the transmission coefficients for a single plant, a_1, and for an array of two plants, a_2, etc. have been assumed to be constants regardless of which part of the crown envelope the ray penetrates. This approximation may cause errors, particularly if some regular plant arrangement is present.

In formula (21) it is considered that the probability distribution P_i may depend not only on the projection area of the crown envelope S and on the shape of the projection region, but also on the location of this region on the horizontal plane and on the azimuth ϕ'. Such effects may occur in row-planted stands. So, even if crown envelopes are modeled as bodies of rotation and their projection area does not depend on the azimuth, probabilities P_i might nevertheless depend on the view azimuth.

In addition to the height z and the view direction Ω, the proportion of gaps is determined by the following characteristics: (1) shape and dimensions of an individual plant crown envelope, determining the area S; (2) stand density and the distribution pattern of plants, probabilities P_i; (3) individual plant crown transparency (a_1); (4) mutual shading of plants ($a_2, a_3, \ldots$).

Any type of distribution functions P_i may be used including those measured or simulated on the computer (e.g. Pukkala 1988; Gusakov and Fradkin 1990). Here, a very close connection with the results related to the plant distribution pattern, obtained in quantitative plant ecology (Greig-Smith 1964), should be noted.

Consider that the plants are located independently of each other, so that the probability of counting i plant stems on the area S may be described by the Poisson distribution

$$P_i(S) = (\lambda S)^i \exp(-\lambda S)/i!, \quad i = 1, 2, \ldots, \tag{22}$$

where λ denotes the density of the plant stand, i.e. the mean amount of the plants per unit ground area. From Eqs. (21) and (22) and assuming $a_2 = a_1^2$, $a_3 = a_1^3, \ldots$ we obtain an analogue of Beer's law:

$$a(z, \Omega) = \exp\{-\lambda S(z, \Omega)[1 - a_1(z, \Omega)]\}. \tag{23}$$

The expression in the exponent stands for the mean number of overlaps. In spite of the fact that assumptions of a Poisson distribution (e.g. complete independence of plant locations) should be neglected as unreasonable, this distribution often serves as a good approximation of measured distributions. Moreover, because of its simplicity, the respective gap probability formula leads to some general conclusions on the relations of the gap frequency with the solar zenith angle and structural stand parameters.

Figure 8 shows an example of the angular distribution of gaps in a spruce forest calculated by means of formula (23) together with (11) and (12) and a comparison with the measured gap distribution. Measured values for average tree crown parameters and stand density were used. The crown envelope was modeled as a

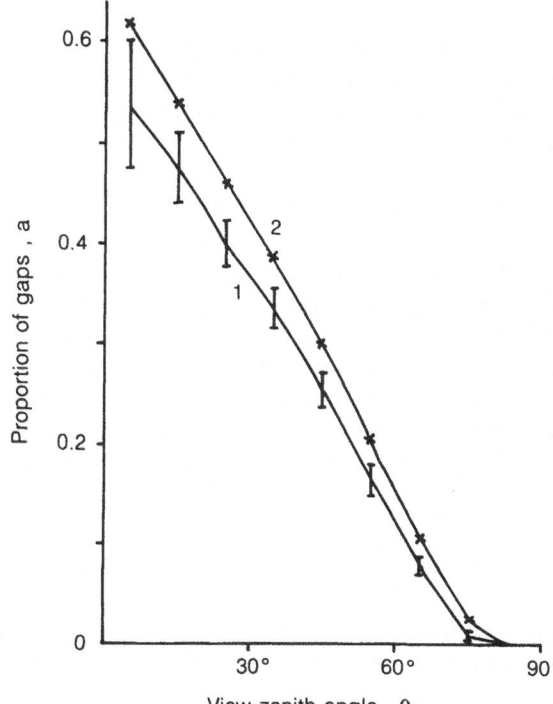

Fig. 8. The proportion of gaps on the floor of an 80-year-old spruce forest as a function of view zenith angle. Curve *1* Estimations from hemispherical photographs (Nilson 1977), *bars* indicate 95% confidence limits of the mean; *2* model calculations for the Poisson-distributed trees

cone in the upper part and a cylinder in the lower part. Since the foliage area index was not measured, the parameter τ was estimated indirectly from formula (12). After Zel'niker (1969) for spruce, the average proportion of gaps in the vertical direction $a_1(0)$ was set equal to 0.05. Evidently, setting $a_1(0)$ to a lower value or considering the distribution of tree size [formula (25), below] would guarantee a better fit.

Experience in using formula (23) in forests has shown that quite often the largest discrepancies between calculated and measured values of the gap proportion occur in the near-vertical view directions. This most probably indicates that regularities in the tree distribution pattern appear namely in those plot areas which are comparable to the plant living space, i.e. the vertical projection area of the crown envelope.

To introduce deviations from the Poisson distribution there is a quite rough but logical way to modify formula (23). I have recommended the following correction term in the exponent, which should account for a clustered or regular distribution pattern, if the deviations from the Poisson distribution are not very large:

$$a(z, \Omega) = \exp\left\{-c\lambda S(z, \Omega)[1 - a_1(z, \Omega)]\right\}, \tag{24}$$

where $c = \ln(GI)/(GI - 1)$, $GI = GI(S)$ being Fisher's grouping index, i.e. the ratio between the variance and the mean value of the stem number occurring in the plot area S. In other words, if there is a clustered distribution pattern ($c < 1$), then the effective number of plants that cause shading is less than in the respective Poisson-distributed stand. Alternatively, in a regular case $c > 1$.

One more extension of formula (23) is possible. When deriving formula (21) all plants in the stand were assumed to be identical. Now consider a plant stand consisting of n different size or species classes, each having its own parameters λ_j, $S_j(z, \Omega)$, $a_{1j}(z, \Omega)$. It can be shown that if all the plants were distributed according to the Poisson distribution, then

$$a(z, \Omega) = \exp\left\{ - \sum_{j=1}^{n} \lambda_j S_j(z, \Omega)[1 - a_{1j}(z, \Omega)] \right\}. \tag{25}$$

If needed, a correction factor c_j for nonrandomness may be introduced into formula (25). For practical purposes an application of formula (25) lies in the possibility of estimating the total amount of radiation intercepted by the plants of the j-th class. First, the whole canopy is divided into sufficiently small horizontal layers determined by a set of heights z_k, $k = 0, 1, \ldots, m$, while $z_0 = 0$ and $z_m = H$. As usual, total interception in a layer Δz_k is calculated as a negative derivative of the gap proportion at z_k. As one can easily see, the part of interception by the j-th class is

$$\lambda_j \{ S_j(z_k, \Omega)[1 - a_{1j}(z_k, \Omega)] - S_j(z_{k-1}, \Omega)[1 - a_{1j}(z_{k-1}, \Omega)] \}$$
$$\cdot \exp\left\{ - \sum_{j=1}^{n} \lambda_j S_j(z_k, \Omega)[1 - a_{1j}(z_k, \Omega)] \right\}.$$

The total amount of direct solar radiation intercepted by the plants of the j-th class is obtained by taking the sum over all the layers.

The same method is also applicable to the problem of bidirectional probabilities in plant stands. Assuming a Poisson distribution of plants in the horizontal, the bidirectional probability $Q(z, \Omega, \Omega')$ of two lines-of-sight in directions Ω and Ω' of being non-obscured is calculated as follows:

$$
\begin{aligned}
Q(z, \Omega, \Omega') = \exp[& -\lambda\{ [S(z, \Omega) - S_c(z, \Omega, \Omega')][1 - a_1(z, \Omega)] \\
& + [S(z, \Omega') - S_c(z, \Omega, \Omega')][1 - a_1(z, \Omega')] \\
& + S_c(z, \Omega, \Omega')[1 - Q_1(z, \Omega, \Omega')] \}],
\end{aligned} \tag{26}
$$

where $S(z, \Omega)$ and $S(z, \Omega')$ denote projection areas of the crown envelope for directions Ω and Ω', respectively, and $S_c(z, \Omega, \Omega')$ is the area of the common part of the two projection regions, $Q_1(z, \Omega, \Omega')$ denotes the bidirectional gap probability for a single crown [formula (14)]. The main reason for the statistical dependence of gap probabilities in two directions lies in the finite dimensions of plant crowns and foliage elements. From Eq. (26) one can see that if $S_c = 0$, then $Q(z, \Omega, \Omega') = a(z, \Omega) \cdot a(z, \Omega')$, i.e. gap probabilities in these directions are independent. However, the common projection area S_c cannot be zero, if $H > z > H - h$, H being the canopy height and h, the crown length.

4.2 Radiation Scattering

When trying to handle the problems of radiation scattering within nonhomogeneous plant canopies, we are faced with the same difficult problems associated with a single plant. At present, the following solutions to the problem seem to be possible:

1. Introduction of all the horizontal inhomogeneities into the coefficients G and Γ and into the source term of the 3-D transfer equation and solving the equation;
2. Using of the Monte Carlo method;
3. Application of Anisimov–Menzhulin's (1983) method of fluctuations;
4. Application of Norman and Welles' (1983) method of equivalent homogeneous canopies;
5. Various approximations, e.g. the method of statistical linearization,

All the methods, particularly 1–4, lead to laborious calculations. The most universal Monte Carlo method (e.g. Szwarcbaum and Shaviv 1976; Oikawa 1977; Ross and Marshak 1987) has practically no limitations to the stand structure involved. However, the Monte Carlo method is a method most convenient for fundamental research, not for routine calculations.

Anisimov and Menzhulin (1983) and Menzhulin et al. (1987) have proposed a new approach for radiative transfer in nonhomogeneous plant canopies. They allowed the radiation extinction coefficient G and the scattering phase function Γ to fluctuate spatially. As a result, the intensities of the radiation field, I, will fluctuate, too. Representing in the transfer equation intensities I and functions G and Γ as the sum of the respective mean value and fluctuation, Anisimov and Menzhulin derived equations for the mean value and the fluctuations. Applying the Friedman–Keller procedure, used in hydrodynamics, they derived equations for the moments of I and correlation moments. However, numerical calculations were performed for the case when only the extinction coefficient G was varied. Thus, the fluctuations of radiation scattering were ignored. The comparison with Monte Carlo calculations showed a close agreement between the two methods in predicting the mean values, variance and skewness of the radiation penetration coefficient.

A way to treat the radiation scattering problem in nonhomogeneous canopies might be simply to ignore the inhomogeneities and consider the scattering problem for a homogeneous canopy having the same parameters (G, Γ, foliage area index) as the nonhomogeneous canopy on the average. In principle, because of nonlinear relationships, this procedure is incorrect. To consider clustering, evidently, the foliage area density and/or the scattering coefficients for a single foliage element should be respectively diminished.

Norman and Welles (1983) proposed an interesting method to reduce the scattering calculations in the presence of nonhomogeneity to that of an equivalent homogeneous canopy. For each point of interest in the nonhomogeneous canopy, they determined the diffuse radiation penetration coefficient both for the upper and lower hemishperes. To calculate the scattered radiation field they chose an equivalent homogeneous canopy and an appropriate depth in it so that the diffuse radiation penetration coefficients were equal to those determined earlier. This method allows for scattered radiation field variations not only in the vertical coordinate, but also in the horizontal. Although not entirely correct, this method may be treated as a good approximation. However, for each point of interest we have to solve the transfer problem in the equivalent homogeneous canopy.

So, we are in need of simplified, more practical methods. The method of statistical linearization may be effectively used in those cases when analytical formulas of radiation characteristics are available. This is the situation for instance, if the reflectance factor, albedo or the downward scattered flux density is needed and we have the respective (approximate) analytical expression originally derived for a homogeneous canopy. Denote this expression by $y = f(w)$, where w is a structural or optical stand parameter, e.g. the foliage area index. Now treat this parameter as a random variable. Assume further that the expression $y = f(w)$ can be taken as locally valid. This means that variations of the parameter w must be of relatively low frequency in the horizontal. If the function f were linear, then the mean value of y would be $f(\bar{w})$. Non linearity may be approximately accounted for by using the following formula:

$$\bar{y} \approx f(\bar{w})[1 + \sigma_w^2 f''(\bar{w})/2], \tag{27}$$

where σ_w^2 is the variance of the parameter w and f'' denotes the second derivative of f. This formula can be obtained by using the Taylorian series expansion of the function f at the point $\bar{w}$. The correction for nonlinearity may be less than unity if $f'' < 0$, and more than unity if $f'' > 0$. Nilson (1991) has used this approach to account for the variable leaf area index in multiple scattering reflectance calculations. It appeared that variations in the leaf area index always caused a decrease in the multiple scattering reflectance factor in the near-infrared region.

4.3 Plant Canopy Reflectance Models

For modeling purposes, it is useful to consider the canopy reflectance as the sum of three components: (1) the reflectance caused by first-order scattering on plant crowns or subcanopies; (2) the part of reflectance caused by single reflection on the soil (or on ground vegetation in forests); (3) the portion of reflectance caused by multiple scattering in the canopy. Sometimes it is reasonable to divide the third component into two parts: reflection from the canopy and reflection from the soil. Such a division is inspired by analyzing high-resolution canopy photos (e.g. aerial forest photos), where four different brightness levels can be distinguished: sunlit crowns, sunlit soil, shaded crowns, shaded soil. To obtain the total reflectance we must either evaluate all the reflectance components directly or estimate the relative part of each component in the scene and its mean radiance.

Recently, Goel (1988) published an extensive review on canopy reflectance models. Among others, there are several reflectance models dealing with vegetation having horizontal inhomogeneities. According to Goel's classification, these are called geometrical models, hybrid models and computer simulation (Monte Carlo) models. Most of the geometrical models give theoretical estimates for the probability of seeing sunlit parts of crowns and soil and those shaded. The models differ mostly in the crown shape models and plant distribution patterns used. Reflectance in shaded areas, i.e. multiple scattering, is often ignored.

Models categorized by Goel as hybrid models for heterogeneous canopies mostly deal with row crops. Welles and Norman (1990) extended their GAR model to calculate the bidirectional reflectance of the canopy (BIGAR model).

A forest reflectance model (Nilson 1991), also a hybrid model, is based on theoretical estimates for the above four reflectance components. The first-order scattering on tree crowns is modeled by Kuusk's (1987) model for individual tree crown reflectance [formula (18)]. Kuusk's bidirectional gap probability Q_1 in the integral is multiplied by another bidirectional gap probability that both lines-of-sight outside the crown of interest are not intercepted by the foliage of adjacent trees. A bidirectional gap probability, defined for the ground level $z = 0$ together with the reflectance factor of the ground vegetation, forms the second component. Calculations of multiple scattering are based on solving the radiation transfer equation by a rough two-stream approximation.

Calculations by means of this forest reflectance model have indicated that grouping of the foliage into crowns and shoots and the presence of ground vegetation give rise to new types of relations between canopy structural characteristics and canopy reflectance factors. For instance, in agricultural crops, along with an increase in the leaf area index, the canopy reflectance in the near-infrared region increases, too. In forests it might be even just the opposite, if the reflectance of ground vegetation is high.

With models to calculate the angular distribution of the reflectance factor, the calculation of the albedo is reduced to the integration over the hemisphere and over the spectrum, if the integral albedo is needed.

5 Conclusions

The structural and optical parameters that determine the radiation field in homogeneous plant canopies are the foliage area index and its vertical distribution, inclination and azimuthal distribution of foliage elements, foliage reflection and transmission coefficient, and the soil reflectance factor. For penumbra and reflectance calculations, the height of the canopy and a parameter for element size are required, too. In canopies with horizontal inhomogeneities all parameters vary in the horizontal. These variations might be given in the form of regular changes or as a certain random array of subcanopies. In the latter case, a distribution pattern of individual subcanopies as well as the geometrical parameters of subcanopies are needed.

Most of the spatial nonrandomness of foliage can be explained by a model grouping the foliage into subcanopies. Within subcanopies a random (Poisson) dispersion of the foliage elements or smaller subcanopies (not necessarily a uniform distribution) may by assumed. Horizontal inhomogeneities in each hierarchy level always cause an increase in radiation penetration and a decrease in interception.

The existing theoretical concepts enable us to treat different levels of the hierarchy (e.g. shoot, whorl, plant, plant stand) in the same manner. If needed, the influence of each hierarchy level may be estimated separately.

However, there are many problems related to radiative transfer in nonhomogeneous plant canopies that remain unsolved. Presently, we are able to calculate radiation penetration, interception and reflection reasonably well. But the greatest problem is how to determine the input data required. It is very difficult to attempt to measure all the parameters and distribution functions required for nonhomogeneous canopies. Therefore, more models of canopy structure, in particular, models of individual plant structure are needed.

References

Acock B, Thornley JHM, Warren Wilson J (1970) Spatial variation of light in the canopy. In: Šetlik I (ed) Prediction and measurement of photosynthetic productivity. Pudoc, Wageningen, pp 91–102

Allen LH Jr (1974) Model of light penetration into a wide-row crop. Agron J 66:41–47

Anisimov OA, Menzhulin GV (1983) On the statistical laws of radiation transfer within inhomogeneous vegetation. Meteorol Hydrol 7:61–66 (in Russian)

Borel-Donohue CC (1988) Models for backscattering of millimeter waves from vegetation canopies. PhD Thesis, Univ Massachusetts

Brown JK (1978) Weight and density of crowns of Rocky-Mountain conifers. USDA For Serv Res Pap INT-197, p 56

Campbell GS (1986) Extinction coefficients for radiation in plant canopies calculated using an ellipsoidal inclination angle distribution. Argic For Meteorol 36:317–321

Chandrasekhar S (1960) Radiative transfer. Dover, New York

Charles-Edwards DA, Thornley JHM (1973) Light interception by an isolated plant: a simple model. Ann Bot 37:919–928

Charles-Edwards DA, Thorpe MR (1976) Interception of diffuse and direct-beam radiation by a hedgerow apple orchard. Ann Bot 40:603–613

Davison B (1958) Neutron transport theory. Oxford Univ Press, London

Daynard TB (1971) Characterization of corn (Zea mays L.) canopies from measurements of individual plants. Agron J 63:133–135

Denholm JV (1981) The influence of penumbra on canopy photosynthesis. I. Theoretical considerations. Agric Meteorol 25:145–161

Dylis NV, Nosova LM (1977) Phytomass in forest communities near Moscow. Nauka, Moscow, p 143 (in Russian)

Fukai S, Loomis RS (1976) Leaf display and light environments in row-planted cotton communities. Agric Meteorol 17:353–379

Goel NS (1988) Models of vegetation canopy reflectance and their use in estimation of biophysical parameters from reflectance data. Remote Sens Rev 4:1–212

Goel NS, Strebel DE (1984) Simple beta distribution representation of leaf orientation in vegetation canopies. Agron J 76:800–802

Greig-Smith P (1964) Quantitative plant ecology (2nd edn). Butterworth, London

Gusakov SV, Fradkin AI (1990) Computer modeling of spatial structure of forest communities. Minsk, Nauka i Tekhnika, p 112 (in Russian)

Hari P, Kaipiainen L, Korpilahti E, Mäkelä A, Nilson T, Oker-Blom P, Ross J, Salminen R (1985) Structure, radiation and photosynthetic production in coniferous stands. Univ Helsinki,, Dept Silvic, Res Notes 54, p 233

Kanevski VA, Ross JK (1985) Effect of the architecture of a conifer on directional distribution of its reflectance: a Monte-Carlo simulation. Sov J Remote Sens 3:659–663 (A cover to cover translation)

Kimes DS, Kirchner JA (1982) Radiative transfer model for heterogeneous 3-D scenes. Appl Opt 21:4119–4129

Koppel A, Oja T (1984) Regime of diffuse solar radiation in an individual Norway spruce (Picea abies (L.) Karst.) crown. Photosynthetica 18:529–535

Kuuluvainen T, Pukkala T (1987) Effect of crown shape and tree distribution on the spatial distribution of shade. Argic For Meteorol 40:215–231

Kuuluvainen T, Kanninen M, Salmi J-P (1988) Tree architecture in young Scots pine: properties, spatial distribution and relationships of components of tree architecture. Silva Fennica 22:147–161

Kuusk A (1985) The hot spot effect of a uniform vegetative cover. Sov J Remote Sens 3:645–658 (A cover to cover translation)

Kuusk AE (1987) Scattering of direct solar radiation in the crown of an isolated tree. Earth Res Space 2:106–111 (in Russian)

Lemeur RL, Blad BL (1974) A critical review of light models for estimating the shortwave radiation regime of plant canopies. Agric Meteorol 14:22–52

Li X, Strahler AH (1988) Modeling the gap probability of a discontinuous vegetation canopy. IEEE Trans Geosci Remote Sens 26:161–170

Loomis RS, Williams WA, Duncan WG (1967) Community architecture and the productivity of terrestrial plant communities. In: Pietro AS, Green FA, Army TJ (eds) Harvesting the sun. Academic Press, New York, pp 191–208

Mann JE, Curry GL, Hartwell DJ, DeMichele DW (1977) A general law for direct sunlight penetration. Math Biosci 34:63–78

Mann JE, Curry GL, Sharpe PJH (1979) Light interception by isolated plants. Agric Meteorol 20:205–214

Menzhulin GV, Koval LA, Anisimov OA (1987) Methods of statistical theory of the radiation regime in case of nonhomogeneous canopy. Trans State Hydrol Inst, Leningrad 327:44–60 (in Russian)

Monsi M, Saeki T (1953) Über den Lichtfaktor in den Pflanzengesellschaften und seine Bedeutung für die Stoffproduktion. Jpn J Bot 14:22–52

Myneni RB, Impens I (1985) A procedural approach for studying the radiation regime of infinite and truncated foliage spaces. Theoretical considerations. Agric For Meteorol 33:323–337

Myneni RB, Ross J (eds) (1991) Photon-vegetation interactions: applications in optical remote sensing and plant ecology. Springer, Berlin Heidelberg New York

Myneni RB, Ross J, Asrar G (1989) A review on the theory on photon transport in leaf canopies. Agric For Meteorol 45:1–153

Nichiporovich AA (1961) Characterization of plant canopies as optical systems. Fiz Rast 8:536–546 (in Russian)

Niilisk H, Nilson T, Ross J (1970) Radiation in plant canopies and its measurement. In: Setlik I (ed) Prediction and measurement of photosynthetic productivity. Pudoc, Wageningen, pp 165–177

Nilson T (1971) A theoretical analysis of the frequency of gaps in plant stands. Agric Meteorol 8:25–38

Nilson T (1977) A theory of radiation penetrantion into non-homogeneous plant canopies. In: The penetrantion of solar radiation into plant canopies. Estonian Acad Sci, Inst Phys Astron, Tartu, pp 5–70 (in Russian)

Nilson T (1984) Calculation of the radiation regime of a forest. In: Radiative climatology and applied actinometry. Materials XII Conf Actinometry, Irkutsk, pp 179–181 (in Russian)

Nilson T (1990) A forest canopy reflectance model. Earth Res Space 3:63–72 (in Russian)

Nilson T (1991) Approximate analytical methods for calculating the reflection functions of leaf canopies in remote sensing applications. In: Myneni RB, Ross J (eds) Photon-vegetation interactions: applications in optical remote sensing and plant ecology. Springer, Berlin Heidelberg New York, pp 161–190

Nilson T, Kuusk A (1985) Approximate analytic relationships for the reflectance of agricultural vegetation canopies. Sov J Remote Sens 4:814–826 (A cover to cover translation)

Nilson T, Kuusk A (1989) A reflectance model for the homogeneous plant canopy and its inversion. Remote Sens Environ 27:157–167

Nilson T, Ross V, Ross J (1977) Some problems on the architecture of plants and plant canopies. In: The penetration of solar radiation into plant canopies. Estonian Acad Sci, Inst Phys Astron, Tartu, pp 71–144 (in Russian)

Norman JM (1975) Radiative transfer in vegetation. In: de Vries DA, Afgan NH (eds) Heat and mass transfer in the biosphere. Part 1. Transfer processes in the plant environment. Wiley, New York, pp 187–205

Norman JM, Jarvis PG (1975) Photosynthesis in sitka spruce (*Picea Sitchensis* (Bong.) Carr.). Part V. Radiation penetration and a test case. J Appl Ecol 12:839–878

Norman JM, Welles JM (1983) Radiative transfer in an array of canopies. Agron J 75:481–488

Norman JM, Miller EE, Tanner CB (1971) Light intensity and sunfleck-size distribution in plant canopies. Agron J 63:743–748

Oikawa T (1977) Light regime in relation to plant population geometry II. Light penetration in square-planted populations. Bot Mag 90:11–22

Oker-Blom P (1984) Penumbral effects of within-plant shading on radiation distribution and leaf photosynthesis: a Monte-Carlo simulation. Photosynthetica 18:522–528

Oker-Blom P (1986) Irradiance distribution and photosynthesis of a Scots pine shoot as influenced by shoot structure and solar radiation field geometry. In: Fujimori T, Whitehead D (eds) Crown and canopy structure in relation to productivity. Forestry and Forest Products Research Institute, Ibaraki, Japan, pp 382–395

Oker-Blom P, Kellomäki S (1982) Theoretical computations on the role of crown shape in the absorption of light by forest trees. Math Biosci 59:291–311

Oker-Blom P, Kellomäki S (1983) Effect of grouping of foliage on the within-stand and within-crown light regime: comparison of random and grouping canopy models. Agric Meteorol 28:143–155

Oker-Blom P, Smolander H (1988) The ratio of shoot silhouette area to total needle area in Scots pine. For Sci 34:894–905

Oker-Blom P, Kotisaari A, Kellomäki S, Ross J, Smolander H (1986) Crown projection area of young *Pinus sylvestris*: a model and its test. Scand J For Res 1:67–74

Philip JR (1965) The distribution of foliage density of single plants. Aust J Bot 13:411–418

Pukkala T (1988) Effect of spatial distribution of trees on the volume increment of a young Scots pine stand. Silva Fennica 22:1–17

Ross J (1964) Photosynthetically active radiation (PAR) in plant stands and its mathematical modeling. In: Actinometry and atmospheric optics. Nauka, Moscow, pp 251–256 (in Russian)

Ross J (1972) Theory of direct solar radiation in horizontally nonhomogeneous plant canopies. In: Solar radiation and productivity of plant canopies. Inst Phys Astron, Estonian Acad Sci, Tartu, pp 122–147 (in Russian)

Ross J (1981) The radiation regime and the architecture of plant stands. Junk, The Netherlands

Ross JK, Marshak AL (1987) Estimation by means of Monte-Carlo method the influence of the canopy architecture parameters on the bidirectional reflectance. Earth Res Space 4:86–93 (in Russian)

Ross J, Nilson T (1965) Penetration of direct solar radiation in agricultural crops. In: Problems of radiation regime of plant canopies. Inst Phys Astron, Estonian Acad Sci, Tartu, pp 25–64 (in Russian)

Shell GSG, Lang ARG (1975) Description of leaf orientation and heliotropic response of sunflower using directional statistics. Agric Meteorol 15:33–48

Szwarcbaum I, Shaviv G (1976) A Monte-Carlo model for the radiation field in plant canopies. Agric Meteorol 17:333–352

Vanderbilt VC, Grant L (1985) Plant canopy specular reflectance model. IEEE Trans Geosci Remote Sens 23:722–730

Vygodskaya NN (1981) Solar radiation regime and structure of mountain forests. Gidromet-izdat, Leningrad, p 261 (in Russian)

Warren Wilson J (1965) Point quadrat analysis of foliage distribution for plants growing single or in rows. Aust J Bot 13:405–409

Welles JM, Norman JM (1991) Photon transport in discontinuous canopies: a weighted random approach. In: Myneni RB, Ross J (eds) Photon-vegetation interactions: applications in optical remote sensing and plant ecology. Springer, Berlin Heidelberg New York, pp 389–414

Whitfield DM (1986) A simple model of light penetration into row crops. Agric For Meteorol 36:297–315

Wit de CT (1965) Photosynthesis of leaf canopies. Agric Res Rep 663, Pudoc, Wageningen, The Netherlands

Zel'niker YuL (1969) Radiation regime under forest canopy. Nauka, Moscow, 98 pp (in Russian)

Deforestation, Revegetation, Water Balance and Climate: An Optimistic Path Through the Plausible, Impracticable and the Controversial

E.A.N. GREENWOOD

Contents

1 Introduction

Deforestation is an intensely controversial global issue. It affects us all. Some call for deforestation to cease on atmospheric, climatic and conservation grounds. Others, particularly the hungry, see in deforestation a way to grow more food. Agencies and resource managers seek to apply accurate information for decision making, yet where is that information to be found? The scientific literature reveals the seeming impossibility of producing rigorous research which can be widely applied to weighing damage against benefit.

Who then can review such a range of issues with authority? Those who seem masters of particular components may themselves be challanged. Reifsnyder (1989) in his lucid critique *A Tale of Ten Fallacies: The Skeptical Enquirer's View of the Carbon Dioxide/Climate Controversy* challenges current proclamations about climate changes.

The timely reviews of Allaway and Cox (1989) and Hamilton (1987) bring rationality to the issues of forests, competing land uses and the consequences for river basins. Can hydrology and related disciplines produce the data to apply to that rationale?

I start with a brief critique of water balance hydrology to assess the rigour of concepts and the accuracy and practicability of measurements. We become involved with climatology, meteorology, hydrology, soil physics, geophysics and hydrogeology, mathematics, modelling, and the biologic sciences. With regret I ignore here the crucial contribution now coming to water balance issues from the socioeconomic sciences, particularly in decision making and public perception.

My expertise is in vegetation hydrology and plant physiology. Experts in other fields must bear with me as I rely more completely on their literature. I know of no earlier review which covers the wide area to which I have been assigned.

The term water balance is a self-evident concept of beguiling simplicity. It implies that precipitation (rain, fog, snow) falling over a catchment or watershed can be totally accounted for in its different states and pathways. These are water vapour from evaporation and water in runoff, in soil, in aquifers, in discharge to streams and to leakage.

Historically, the water balance concept has been invoked to estimate total evaporation which, traditionally, has been considered the most difficult term to measure. If the values of precipitation and the storage and drainage terms are known accurately, then total evaporation can be derived. And the reason why evaporation is required is its close connection with vegetation management. Evaporation can be manipulated easily by modifying the vegetation-deforestation, afforestation, reforestation, cropping, or pasture. If total evaporation could be measured directly and conveniently, then there would be less need to invoke water balance.

The tragedy for hydrology is that, for decades, the water balance approach was taken in the hope that it would work. Sometimes water budgeting did balance in small catchments (e.g. McIlroy and Dunin 1982). Even in such cases it might not be rigorously demonstrated that the balance is other than a coincidence of sampling errors.

The critique is followed by a review of the influence of site operations on water balance. The main section deals with the reported effects of deforestation and revegetation on water balance. There is a concluding justification for optimism.

As this is an inter disciplinary work, any one reader may not be familiar with all of the concepts and terminology. Accordingly, I give some simple introductions which may seem naive to experts. I also avoid symbols and equations.

2 A Critique of Water Balance

Hydrologists face a dilemma in assessing the effects of changes in vegetation on water balance. At what scale should they operate? Small-scale studies on selected parts of a catchment can reveal pathways, rates of processes, special features, and variability. Whole-catchment investigations can provide only bulk values from which one is likely to learn about broad outcomes rather than processes. The dilemma is acute for modellers.

To the extent that measurement is rigorous, we may deem it valid for the experimental site. But to what extent does that site (or sites) apply to the rest of the catchment? It is here that rigour is knowingly lost but one does not know how serious the error may be.

At the whole-catchment scale, the increase in yield of surface water from given deforestation treatments may be accurately measured. But nothing is known of the influence those treatments had on other components and pathways of water balance. Nor may one learn what the water yield (or quality) would have been if other logging methods or areas had been used. Such research is of undoubted value but it is "blunt".

The literature of hydrology, and water balance in particular, contains what might be called assumptions of desperation. They seem to be generally accepted, albeit reluctantly, by both authors and editors. A few examples of published assumptions are:

* That there are no leaks in the catchment;
* There is uniformity between two similar, adjacent drill cores;
* Catchment divide corresponds with topography;
* Preferred pathways or barriers below ground which are not visible do not exist;
* Forest understorey does not contribute greatly to total evaporation;
* Transpiration is entirely physically (or energy) driven;
* A single (or a few) evaporation pan or rain gauge adequately represents the catchment.

Such assumptions add fuel to the burning indictment of dilettantism in hydrology (Klemeš 1986). Modellers are in the worst position. They must take process data which have been inadequately sampled and measured and from them derive extrapolations, predictions and optimizations. On the water balance equation, Klemeš (1988), in his Presidential Paper, remarks on the sometimes poor quality of thinking and expertise that has been published. Readers, whatever their discipline, should read that paper and ask of themselves whether some of his criticism applies beyond hydrology. Yet we need not despair; some splendid, recent advances justify optimism.

Beyond that criticism is finance. I expect that all researchers are disappointed that they have insufficient resources to support the extent and rigour which they know their projects require.

2.1 Precipitation

Precipitation, usually rain, snow or fog, is the basic parameter of terrestrial water balance. It has often been considered to be the easiest to measure accurately. But is that so?

2.1.1 Rain

The inaccuracy of the standard rain gauge was found by Allerup and Madsen (1980) to be about -15%. The shortfall is due manily to wind exposure and a little to wetting loss. They devised an accurate correction model which used meteorological data from automatic weather stations. Kay (1984) makes important observations on gauge performance: (1) shortfall increases with low intensity events and varies with exposure; (2) extreme rainfall combined with high wind causes serious error in flood forecasting and design analysis; (3) wind direction markedly influences gauge exposure in clearings and mountainous areas; and (4) exposure is influenced by the seasonal cycle in deciduous forests, frozen snow on conifers, and by disease, insects and fire.

How far apart should rain gauges be spaced? There are meteorologic and terrain factors to be considered. Tropical rainfall with its extreme local variation requires a closer spacing of gauges and the more so if the land use was small-scale peasant agriculture as against forests (Jackson 1978).

There are severe and costly constraints in measuring rain in large and remote catchments of difficult terrain, e.g. installation, servicing and providing automatic and instantaneous (for flood forecasting) transmission of data. In this respect, Collier (1986a–c) examined the accuracy of estimates of rainfall by radar. By introducing an interactive calibration of unmanned radar by five automatic rain gauges, he increased the accuracy of hourly rainfall within 75 km of the radar site. Collier's discussion of the practical issues and of further research in remote areas is challenging reading.

Lebel et al. (1987) sought to improve the accuracy of the rainfall field, as distinct from point, values of rainfall. They started with a high sample density and explored loss in accuracy from sparser sampling. Then they used the scaled estimation error variance as computed from a scaled climatologic variogram model of the rainfall field. The validated error variance is then used to assess the performance of three different linear estimators – Thiessen polygons, spline surface fittings and kriging – with varying network densities.

Other approaches to areal rainfall modelling have been made. Rodriguez-Iturbe (1986) dealt with the scale of fluctuation of rainfall models in a way which is helpful to hydrologists needing an average rainfall over a certain period of time. Obeysekera et al. (1987), while comparing three rainfall models in Colorado, observed that diurnal periodicity of storms predominates in summer, which is a feature not built into any of the models used!

Accurate areal rainfall data may be expensive to obtain; not to acquire it is to incur expensive errors.

2.1.2 Snow and Ice

With respect to water balance, the water equivalent of snow may seem to be a sufficient quantification. Yet the speed with which snow melts determines, in part, the partitioning into evaporation, runoff and infiltration. The rate of infiltration depends on the extent to which the soil is frozen; the runoff is influenced by available energy and the state in which the snow is deposited, both aspects being influenced by the nature of vegetation cover.

Burt and Williams (1976) developed a special permeameter for measuring the hydraulic conductivity of frozen soils which may contain substantial quantities of water (Fig. 1). Kane and Stein (1983) give a lucid account of the complexity, i.e. the effects of degree saturation, the extent of freezing and the degree of fracture of the soil. The accuracy with which snowmelt can be modelled requires an understanding of the behaviour of frozen soils. Burn and Smith (1985) dispute these findings, in

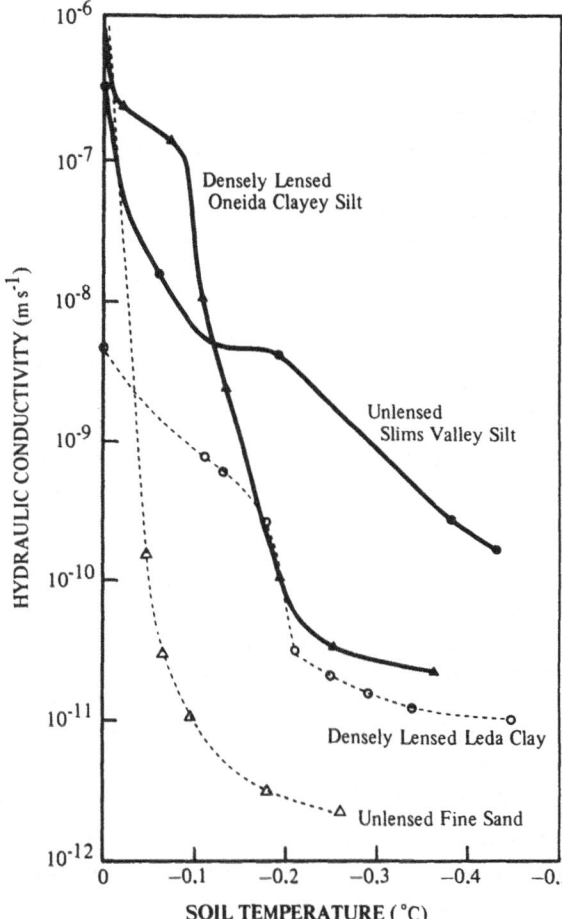

Fig. 1. The influence of soil type on the hydraulic conductivity of frozen soils (after Burt and Williams 1976)

particular, the determination of soil moisture potential gradients and their role in calculating the hydraulic conductivity of frozen soil.

Given that the processes are understood, a wide range of infiltration rates in frozen soils occurs, which, in turn, induces variation in the amount of snowmelt runoff. The consequence for water balance is that, in practice, snowmelt over frozen soil cannot be determined accurately. The difficulty of rigorously measuring snowfall, or its water equivalent, as well as the effects of vegetation is described by Krutilla et al. (1983), by Golding and Swanson (1986) who used energy balance analyses, and by Blackie (1987). The difficulties seem insuperable. And, as a large proportion of terrestrial precipitation is as snow, snowmelt becomes a most important event in the hydrologic year. A critical advance in resolving this issue has been the use of "environmental" isotopes, deuterium and oxygen-18, to separate "event" water (snowmelt and rain) from "pre-event" water (Buttle 1989).

The nature of vegetation cover has a sensitive influence on the behaviour of snow. Lusby (1979) states that the minor change from sagebrush to grass could increase runoff from spring snowmelt by 12% caused, it is claimed, by a change in snowpack. Caprio et al. (1989a), using lysimeters, found that simply leaving stubble standing effectively increased soil water by increasing the snow entrapment. It was also sensitive to wind (chinook) strength (Caprio et al. 1989b). What then might we expect from forest clearing?

When an aspen forest was clearcut (Verry et al. 1983), annual snowmelt peaks doubled and occurred 5 days sooner (although snowmelt runoff volume did not change).

Even greater effects on snowmelt occur when non-deciduous forests are cleared (Harr 1986; Berris and Harr 1987). The effects of deforestation and revegetation on snowmelt is presented in Sect. 4.1.2.

Several models have been developed recently to use snowmelt parameters in water balance. Kattelmann et al. (1985) estimate snow water equivalent; Haltiner and Salas (1988) provide short-term forecasting of snowmelt runoff; Harding (1986) models exchange of energy and mass in melting snowpeak; Taniguchi and Kayane (1986) deal with changes in soil temperature caused by infiltration of snowmelt water and Hoggan et al. (1987) have published a real-time snow simulation model.

Buttle and McDonnell (1987) discuss the most effective parameters for modelling. For well-exposed areas possessing a discontinuous, shallow snowpack, a model should be used which assumes that melt takes place at the margins; for deep areas, a uniform melt is preferred. Applying these criteria to models 4 and 5 of Dunne and Leopold (1978), very accurate simulations are obtained.

Precipitation as snow, and the complications of ice, are still in practice semi-quantitative parameters of water balance. Predicting or quantifying snowmelt runoff over the wide range of physiographic and climatic conditions, and forest management options which are available, are major issues for international research. Those concerned with snow should not miss reading the outstanding workshop paper by Leavesley (1989). He identifies the major issues common to all regions:

* Definition of spatial and temporal distribution of model input;
* Measurement or estimation of snow accumulation, snowmelt, and runoff process parameters for a range of application and scales;
* Development of accurate short- and long-term snowmelt runoff forecasts.

Procedures being developed to solve those issues are:

1. Integrating conventional and remote sensing data to improve estimates of input data;
2. Developing snowmelt process algorithms that are closely related to measurable basin and climatic characteristics;
3. Updating model parameters and components using measured data or knowledge of past uncertainty;
4. Development of improved model capabilities and establishment of standardized techniques and measures to evaluate model performance and results.

The importance of Leavesley's paper lies in the quality of analysis and proposals for the above issues.

There is then much work to be done on snow and ice to attain satisfactory water balance standards. Two techniques have been applied to make basic measurements more accurate. The first is the explicit measurement of numbers of snow particles and the use of terrestrial gamma radiation to estimate the snow water equivalent. Schmidt and Troendle (1989) compared the number flux of snow particles in forests and clearings 80 m wide. Their conclusions were:

1. In light winds, the number flux of precipitation particles decreased significantly at upper levels within the canopy of mature spruce-fir forest, compared with the flux above the canopy.
2. Similar differences occurred between the levels in the centre of a clearing of size four times the forest height.
3. Number fluxes above the centre of the clearing were greater than those measured 10 m above the upwind forest canopy.
4. Wind speed in the range $0-6\,\mathrm{m\,s}^{-1}$ appeared to affect directly the difference between forest and clearing fluxes.

The authors are concerned that mass flux $(\mathrm{g\,s}^{-1}\,\mathrm{cm}^{-2})$ would be preferable to number flux (number concentration x wind speed). For future work they are developing a mass flux instrument.

The second technique is the estimation of the snow water equivalent by terrestrial gamma radiation (Carroll and Carroll 1989). If variability of snowpack exists along a flight line, estimates are systematically underestimated. Three expressions, which depend on the variance and the shape of the snow water equivalent distribution, are derived to estimate the size of the error in the airborne estimate. Results from two flights are presented.

2.1.3 Fog Drip or Mist Precipitation

In his review of mist precipitation on vegetation, Kerfoot (1968) concluded that, while the quantity could be substantial, measurement was inaccurate. How far have we advanced since then?

A major advance was Shuttleworth's (1977) analysis of the exchange of water between wind-driven fog and natural vegetation, i.e. the direct capture of fog droplets and the interaction between evaporation and condensation. The analysis has strong implications for the method of measurement of mist precipitation and for the likely influence of vegetation types (including deforestation/afforestation).

O'Connell and O'Shaughnessy (1975) and Harr (1982) used troughs in forested and cleared land to evaluate fog drip. Such a procedure confounds rainfall interception with fog drip. O'Connell and O'Shaughnessy partially resolved the issue by using only rainless days. Harr used interception data in another year from a similar site (Rothacher 1970) but with little fog occurring. Authors of both studies were aware that their findings were not rigorous.

Ingraham and Matthews (1988) used natural isotopes to measure the importance of fog drip as a source of groundwater recharge. But they too were unable to determine clearly the proportion of recharge attributable to fog drip even though it seemed substantial.

For some regions fog drip may be a substantial component of water balance. We have not advanced far since 1968 in our ability to measure it accurately.

2.1.4 Volume of Precipitation

To obtain the volume of precipitation, we need to know the area of the catchment. This is usually taken as the area bounded by the topographic divide, whereas for water balance it ought to be the hydrogeologic divide. Leakage into and out of a catchment, and below-ground diversions, induces unknowable errors. Those issues are presented in Sect. 2.5.

The conclusion to be drawn is that all of the components of precipitation are difficult to measure accurately in all but some small experimental first-order catchments.

2.2 Evaporation

The measurement of total evaporation is crucial to the issue of vegetation and water balance. Total evaporation includes the interception of precipitation by foliage, transpiration, and evaporation from the soil surface. Historically, the most important event was the Penman equation (1948) for evaporation. There were several other equations developed subsequently, but the Penman equation is probably still the most widely used estimator of total evaporation for vegetated surfaces by hydrologists. Calder (1990) gives a lucid exposition and evaluation of those several classic equations.

In my opinion (and in the context of this review), evaporation research has been brilliantly profound, voluminous, unnecessarily controversial, obsessed with the physical, and of uncertain rigour in application. Extensive theoretical papers continued to be published throughout the 1980s, but they do move toward the resolution of controversy. The biologists surprised us with evidence of partial non-physical control of transpiration. Instruments improved in accuracy and practicability. Modelling became interdisciplinary and plausible. Yet extensive, rigorous measurement of evaporation continues to be minimal!

Two questions arise:

* Why are the new methods not being used extensively in resolving vegetation and water balance issues?
* How are models of, and equations for, evaporation to be tested in practice?

2.2.1 Interception of Precipitation

The usual method of measuring interception is to place either funnel or trough rain gauges near ground level in adjacent cleared and forested areas. The difference between the mean values in the two sets of gauges represents interception. The three main sources of error are: (1) inadequate sampling in heterogeneous vegetation; (2) variation in rainfall with vegetation density (Sect. 2.1.1); and (3) rain gauge design (Sect. 2.1.1).

In very large catchments, adequate sampling is impossible unless remote sensing (particularly aerial photography) is used to select representative sampling sites. In remote and rough terrain, sampling would never be adequate, although the use of extrapolation models would improve it.

An important application of interception measurement is when new vegetation strategies are being developed following deforestation. What is the best mix of say crop species and trees which provide long-term income without changing water balance too severely from that of the prior forest? We face that issue in Sect. 4. One of the simplest and most powerful counterbalancing agents is interception. Other aspects aside, interception by single trees is maximized by using tall species with a large leaf area and an open canopy.

The following references indicate the range of interception values which occur in different ecosystems. The examples are chosen for their variety and insight.

A critical and extensive partitioning of rainfall into interception, throughfall and stem flow is given in the four-part series of Crockford and Richardson (1990). They compared partitioning in a eucalypt forest and pine plantation in south-eastern Australia over several years (Fig.2). In general, the eucalypt forest intercepted about 11%, and pines about 18%, of precipitation. Management of pines altered the values of interception.

Stewart (1977), in Britain, measured the energy budget of a pine forest on 70 days when the canopy was wet. Out of 245 20-min periods when the canopy was wholly wet, 173 were occasions when the latent heat flux exceeded the net radiation, the additional energy being provided by a downward flux of sensible heat. Under the same level of radiation the avarage rate of evaporation of intercepted precipitation has been found to be three times the average rate of transpiration. It implies that the loss by evaporation of intercepted precipitation is only partly compensated by the suppression of transpiration.

Prebble and Stirk (1980) obtained annual interception rates *beneath* individual trees of *Eucalyptus melanophloia* of about 11% of rainfall (718 mm). The ecosystem was a grassy woodland, so the overall interception would be less. Dunin et al. (1988), using a 40-tonne lysimeter in a dense regrowth forest, found interception to range from 10–15%, over 3 years, depending on the wide fluctuation in annual rainfall

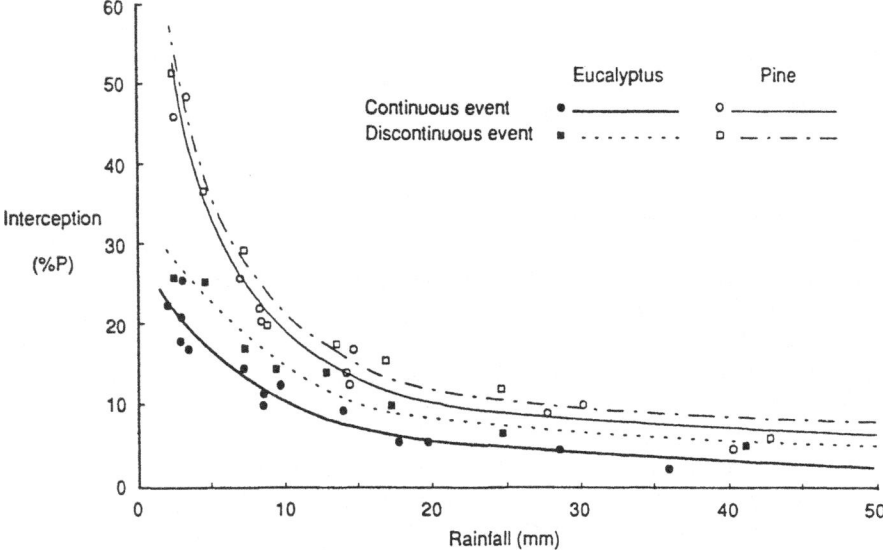

Fig. 2. The interception of rainfall by eucalypt forest and pine plantation as affected by magnitude and duration of rain in south-eastern Australia. Interception is expressed as a percentage of precipitation (After Crockford and Richardson 1990)

$(600-1500 \, \text{mm yr}^{-1})$. Smith (1974) compared a pine plantation with a eucalypt forest (860-mm rainfall). Pines intercepted 19% compared to 11% by eucalypts.

Greenwood et al. (1985a) compared interception by several species of 9-year-old eucalypts in small, and thereby, advective plantations with trees 3.5 m apart. Interception ranged from 16–37% of rainfall $(800 \, \text{mm yr}^{-1})$. The climate was Mediterranean.

Rutter (1963) obtained interception values of 32% in *Pinus silvestris* in a 700-mm rainfall. Gash et al. (1980), using coniferous forests, compared measured values of interception with estimates from the Rutter model (Rutter et al. 1971, 1975). Discrepancies were about 20%. The measured values were 27, 42 and 32% at three different sites.

Lloyd and Marques (1988) found interception to be about 7% of rainfall in the Amazonian rainforest. They also made the useful observation that regular, random relocation of gauges was better than fixed positions and that location along a transect was more effective than over an area.

Stogsdill et al. (1989) investigated the effect of a range of precommercial thinning of *Pinus taeda* on throughfall (throughfall = rainfall – interception – stem flow). A prediction equation relating throughfall to basal area and gross rainfall for individual storms was developed. Throughfall data collected periodically during 1986 were used to test the prediction equation. Throughfall during the growing season increased by approximately 3% of total rainfall for every $4 \, \text{m}^2 \, \text{ha}^{-1}$ reduction in basal area. Depending upon the role of stemflow, additional input to the forest floor from throughfall, when reducing the basal area from 26.6 to $12.6 \, \text{m}^2 \, \text{ha}^{-1}$, would range from 22 to 60 mm. This additional water would increase the amount of soil moisture available to the stand during the growing season. The R^2 of the prediction equation was 0.99 and the inclusion of basal area in the equation was significant at $P = 0.001$.

In the West Java tropical rainforest, Calder et al. (1986) estimated interception using a Rutter-type model to be 21% of a rainfall of 2800 mm and total evaporation to be 1480 mm. The latent heat requirement for this evaporation was identical to the net radiation input.

How accurate are simple interception methods? The series of papers by Crockford and Richardson (1990) critically assesses accuracy in a way which allows us to estimate the necessary trade-off between accuracy and scale of operations for a given project.

There are several models for interception of varying complexity. Calder's (1986c) model takes into account mean raindrop volume as an explicit input variable and predicts that, for the same total amounts of rainfall, maximum canopy storage will be attained less rapidly for drops of larger volume.

Since it is unsatisfactory to describe interception models briefly, I shall deal with them by posing two questions:

* Is it better to measure rather than model interception?
* Would it be more efficient to ignore interception and to put one's resources into measuring total evaporation (interception + soil surface evaporation + transpiration)?

Dolman (1987) helps us to answer these questions. He compared an analytical model (Gash 1979) and a numerical simulation model (Mulder 1985) in estimating summer and winter interception by a deciduous forest. The models agreed well. The parameters given in Table 1 were required. Although the requirement for data differs between these two models, both require more data collection than the direct measurement of interception. And that would hold regardless of the area to be sampled.

Table 1. Main differences between the numerical simulation model of Mulder (1985) and the analytical model of Gash (1979) (After Dolman 1987)

Parameter	Gash	Mulder
Rainfall	Daily values of gross precipitation	Daily values of gross precipitation, number of showers, duration of rainfall
Meteorological variables	Mean evaporation rate, mean rainfall rate	Daily mean values of net radiation, temperature, vapour pressure deficit, windspeed
Vegetation parameters	Saturation storage capacity, coefficient of free throughfall	Saturation storage capacity, coefficient of free throughfall, zero plane displacement, roughness length
Time scale	Day	Day

2.2.2 Transpiration

The inability to measure transpiration has been an incentive for water balance studies. To the hydrologist, the other components seemed easier to measure. Transpiration was estimated as a residual. Bearing in mind the inaccuracies of water balance methods in practice, there was usually little confidence in those estimates of transpiration.

The models of transpiration, developed to substitute for direct measurement, were energy-based. Stomates were the supreme regulators, the unseen roots were ignored. In a land-mark review, Turner (1986) exposed the role of roots in influencing the water relations of plants to those who were unaware of current research in plant physiology and biochemistry. Awareness of this field was further heightened by outstanding "opinion and response" articles in *Plant, Cell and Environment* between Kramer (1988), Passioura (1988) and Schulze et al. (1988) on "changing concepts regarding plant water relations".

Another outstanding exposition, presenting bioclimatology at its best, is the analysis by Jarvis and McNaughton (1986) of the differing viewpoints of plant physiologists and meteorologists on the stomatal control of transpiration. They take us from the single stoma to the single leaf, the plant, the canopy, the region and the planetary boundary layer.

Given the fresh insights above, how much closer are we to measuring or modelling transpiration over a range of scale and vegetation type? As the scale increases, process research must give way to inventory. Would it be more effective to abandon transpiration alone and measure total evaporation (Sect 2.2.3)? The answer lies partly with the practicability (and cost) of automatic measurement of transpiration.

Automatic and accurate measurement of transpiration by trees and shrubs using heat pulse techniques is now a practical reality. The volume flux of water through the trunk or stem, below the lowest branch, is obtained. The current main versions might be called the Israeli (Cohen et al. 1981, 1985), the European (Čermák et al. 1984; Schulze et al. 1985) and the New Zealand (Edwards and Warwick 1984). A miniature version of the European model for herbaceous plants has also been developed (Fichtner and Schulze 1990). The New Zealand model provides automatic measurement, data processing and printout every 15 min. An advanced prototype for large-scale, multiple-tree applications has been developed by Hatton and Durham (pers. comm.). Although the heat pulse technique requires very careful attention to detail and precision, it does avoid the impracticability and biases of porometry, isotope and chamber methods. There is now no excuse, other than poverty, for not measuring transpiration more extensively in process hydrology.

Dolman (1988) predicted transpiration of an oak forest from the Penman–Monteith equation and a model of canopy conductance based on parameter measurements of stomatal conductance. An upper limit to transpiration of 2.1 mm day^{-1} is introduced to regulate daily water loss. He discusses the physiologic basis of that decision. The model predicts transpiration to within 10% of observed water loss from a lysimeter. Unfortunately, porometers are notoriously laborious because of the excessive sampling of leaves, so we could not expect such a technique to be applicable to more than a very few trees at one time.

Dolman et al. (1988) predicted forest transpiration from hourly climate data from which they calculated surface conductance for a pine forest and transpiration from the Penman–Monteith equation. Observed and predicted values agreed well (conductance within 3% and transpiration within 0.5%).

Transpiration rate alone does not provide all the information about water use needed for water balance. Where is the water being drawn from, and in what proportions? This returns us to roots and their production rate (Kurz and Kimmins 1987) and distribution in the soil. Further, roots will not be active users of water from the saturated zone unless the oxygen content of the water is adequate (a simple instrument is now available for this purpose which is an adaptation of a methane sensor developed by Barber and Briegel 1987).

Finally, the extrapolation of transpiration by trees on a temporal and an areal basis ought now to become more satisfactory because more, and better, data should be available. Measurement of transpiration by herbage or forest undergrowth is still impracticable.

2.2.3 Total Evaporation

The main instruments for estimating total evaporation, as distinct from transpiration, are:

1. Weighing lysimeter, an accurate, all-weather instrument. It provides only a point measurement, is limited to shallow-rooted vegetation, and is cumbersome to install.
2. Bowen ratio instrument which accurately measures humidity gradients above vegetation. It requires smooth terrain to maximize wind fetch. It is ideal for crops and level forest canopies. It integrates over several hectares and is portable.
3. Ventilated chamber with infrared water vapour analyzer. It measures evaporation over a given area (say $20 \, m^2$) of crop or other low vegetation. It is portable and accurate.
4. Eddy correlation, a small, sensitive complex but accurate instrument limited to dry weather.

The methods above were reviewed by Stewart (1984), with particular reference to forests, and by Oliver (1985); a more extensive and up-to-date statement is given by Calder (1990).

How can we assess the accuracy of one of these instruments? The most basic field instrument, the evaporation pan, is influenced by topographic location, surrounding vegetation and method of installation. Bosman (1987) found that Class A pans above bare soil and crushed stone, and without protective screening, evaporated 6 and 19%, respectively, more than that above grass cover. With bare soil and protective screening, pans evaporated 18% less than those without protection. Such sensitivity to bias is transmitted to those models or equations requiring pan evaporation.

Weighing lysimeters also may not be taken for granted. Dugas and Bland (1989) found no bias due to size alone.

McIlroy and Dunin (1982) verified total evaporation by measuring all components water balance (except leakage) in a small, forested catchment. They also installed a 40-tonne lysimeter to measure total evaporation directly. The two estimates agreed. Then they measured total evaporation both by Bowen ratio and by a ventilated chamber over the trees in the lysimeter. Similar values for total evaporation were obtained from all methods (Dunin and Greenwood 1986).

Dunin et al. (1989) mounted a second exercise on a lupin crop using change in soil water content, Bowen ratio, ventilated chambers, and the hand-held infrared meter and porometer procedure. Again, the methods agreed well, except for infrared, which underestimated total evaporation.

Agreement between methods does not imply accuracy, but it does inspire confidence. Is there, in practice, a better approach?

Van Zyl and De Jager (1987a, b) found that the Penman–Monteith equation accurately predicted total evaporation by grass and wheat as measured by lysimeters. This occurred with or without adjustment for atmospheric stability and whether stability was measured over grass or wheat, a reassuring conclusion! They then compared hourly evaporation rates from a Piche evaporimeter in a Stevenson screen, a Piche evaporimeter outside the screen but shaded from direct radiation, and a carborundum evaporimeter similarly shaded. The data were substituted in the Penman–Monteith equation and again tested with the lysimeter. They concluded that the carborundum evaporimeter was the most sensitive, though the index of agreement was similar, and that the model was reliable. It would be useful for this work to be repeated using vegetation of increasing height and roughness (shrubs and trees) to establish the limits of accuracy of the Penman–Monteith equation.

Leith and Solomon (1985) estimated total evaporation over remote areas where only a precipitation network was available. They adapted the simple, long-term model of Turc (1954) in conjunction with the Geostationary Operational Environment Satellite (GOES) digital images.

Turc's model assumes that, if means of components of the water budget are calculated over a sufficiently long period, the variation in storage of both surface and groundwater is negligible in relation to the other components of water balance. It is also assumed that the groundwater component is negligible compared to precipitation and evaporation. The question here is not one of rigour but whether such estimates are better than guessing. The authors judge error to be 20–60% which I equate with "experienced guessing". But they claim that, with a denser gauging network, the outcome would be better. Since a large proportion of the world's catchments are sparsely gauged at best, a better-than-guessing estimate is a goal to be pursued enthusiastically.

Abdulmumin et al. (1987) determined regional total evaporation from the basic surface energy equation as a function of net radiation and sensible heat flux. This requires only the readily available rawinsonde and global solar radiation. Their highest correlation coefficients of predicted against observed (using bimonthly totals of precipitation runoff for total evaporation) were about 0.90. He predicted that remote sensing of albedo would improve accuracy and recommended the method for large-scale ungauged catchments.

DeBruin (1983) developed a suite of well-known evaporation methods for use in the wet and dry seasons in the humid tropics. In particular he makes use of duration of sunshine for locations where the wet season lasts for more than 7 months. He estimated total evaporation for some 60 stations. The value of the paper lies in the emphasis on the evaluation of practicable methods.

Anyadike (1987) takes us to the heart of a practical issue in several climatic zones (tropical rainy, tropical wet and dry, semi-arid tropical and tropical desert) of West Africa where irrigation is being developed as a result of severe droughts. Which evaporation formula is the most accurate and practicable in the absence of meteorologic stations: Penman's, Thornthwaite's or Linacre's? Here is a classic triangular dilemma with serious consequences. The Penman formula is accepted as the best (a concensus judgement), but it fails in practice because the required data are unavailable. Local experience rates the Thornthwaite formula as erroneous (Ayoade 1976). The simplified Penman formula of Linacre (1977) was found to be a good substitute for Penman estimates over 44 widely distributed sites. The Linacre version requires temperature, elevation, latitude and dew point (the last being the only rare item).

Not everybody, even after all these years, agrees on the merits of the above formulae. Pereira and de Camargo (1989) have just published "An analysis of the criticism of Thornthwaite's equation for estimating potential evapotranspiration". They conclude that due consideration must be given, in "oasis" situations, to the fetch necessary to obtain the potential rate condition. They substantiate their point with examples.

Vast areas in semi-arid climates are occupied by sparse woodlands. A special feature of such ecosystems is that items of vegetation can be discerned individually or in clumps by remote sensing. Vegetation cover models have been developed (Running and Coughlin 1988) from which estimates of total evaporation can be derived. And those estimates can be tested rigorously over a range of scale by Bowen ratio instruments (hectares), ventilated chambers ($20 \, m^2$) or thermo-electric heat pulse on individual trees (Hatton and Vertessy 1990). A most promising and intense development of research into the remote sensing of regional evaporation in semi-arid woodlands is evident from such publications as Jupp and Kalma (1990), Jupp et al. (1990) and Walker et al. (1986).

Calder (1990) presents a chapter on the measurement and modelling of evaporation from vegetation under snow and ice.

Direct methods for measuring total evaporation are capable of large-scale application. Yet it may be more effective to construct a model which, at some smaller scale, has been tested for its validity.

Shuttleworth (1988a, b) demonstrated the effectiveness of working in groups with a wide range of specialists, including modellers. This raises important questions. The more complex the exercise, the harder it is for an outsider to examine it rigorously. And with respect to complex models, how might their validity be demonstrated? Who but the modeller has the time to assimilate the software and to test it? Whereas, in direct measurement, the experimental details are described in the journal paper, and the reader can evaluate it readily. Yet complexity exists and we must embrace it as best we can.

Shutteworth (1988b) estimated that about 50% of evaporation in the Amazonian rainforest was recycled within the region. Dickinson (1989) reviews recent projects which measure and model the effects of deforestation on the regional surface climate in Amazonia (Fig. 3). Further to Amazonia, Gat et al. (1985) give a brief dissertation on the local and downwind effects of reduced evaporation and recycling of rain following intensive deforestation.

In environments with high atmospheric advection but low radiation, such as pine forests in hilly terrain in temperate climates, interception may dominate transpiration. Cooper and Lockwood (1987) review the advantages of multilayer models to deal with such circumstances and the several published versions since 1972. They present a revised model. It is a good example of selecting a model for a given, widespread set of conditions.

The "complementary theory" of Bouchet (1963) has become prominent in the 1980s. It postulates that, as available water becomes limiting, there will be corresponding increase in temperature and a reduction in water vapour pressure of the air. This implies that when the actual evaporation increases, the potential evaporation will decrease, and that the properties of the air mass observed within the boundary layer may be used to infer the amount by which actual evaporation falls short of potential evaporation.

The complementary theory as expounded by Morton (1983) is compelling if it is assumed that the control of transpiration is entirely physical. It is less attractive if we consider a partially independent biological control of stomata by the roots as notionally anticipated by Morton. Morton (1985) presents an excellent explanatory paper on complementary theory. How well does it hold for forests with their canopy resistance?

Byrne et al. (1988) tested the theory using some 800 days of rigorous evaporation data from a forest lysimeter and meteorological records. Concurrently, they applied the Penman–Monteith equation with a fixed canopy resistance, and a simple regression on net radiation. They concluded that the theory offers an alternative and plausible formulation of the relationship between meteorological data and regional evaporation. Their error terms, however, suggest that their assessment was generous!

Morton (1984), in one of the finest debates in hydrologic literature, criticized current work on forest evaporation. He deals with the driving forces of transpiration

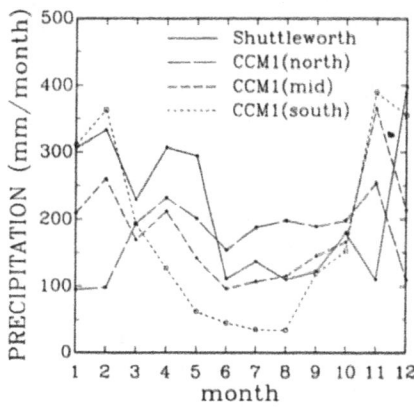

Fig. 3. Comparison of observed (Shuttleworth 1988b) and modelled values of total evaporation at three locations in the Amazonian forest (After Dickinson 1989)

and interception, the reported values of which he was sceptical – they were too high. Calder (1985) challenged this thesis, much of which Morton accepted. In 1989, McNaughton and Spriggs further challenged the validity of the complementary relationship. My abstracts are unworthy of all of these authors so I recommend that you read the originals. Should you do so, you should first read Turner's (1986) review of biological control of stomates.

In 1989, the first four papers in the Journal of Hydrology, Vol. 111, again deal mainly with theoretical aspects of evaporation, but they do resolve those earlier controversies. I expect also that they will lead to more confident measurements in the field, which is urgently required to service the current modelling programs. Those four papers are by Nash (1989), Granger (1989a, b) and Granger and Gray (1989).

Stewart and Gay (1989) present two models of evaporation which they validated with Bowen-ratio instrumentation. Differences between the two models and the measurements were $\pm 5\%$. The vegetation was tall grass prairie.

In order to estimate evaporation from a wide range of complex terrain in Japan, Hoshi et al. (1989) developed a system based on Landsat MSS, elevation and meteorological data. They tested their methods against Thornthwaite's method, pan evaporation and the water balance equation. Only in some seasons did their methods agree well.

Finally, Dunin (1990) presents the issue of scale in the measurement of evaporation. Knowing that most evaporation measurements have been on a small scale, he explores the possibility, and reliability, of extrapolation to regions. He concludes that point source data are quite inadequate, but increased deployment of evaporation measurements over a range of seasonal conditions would greatly facilitate the estimation of regional evaporation.

2.3 Distribution of Water

The distribution of water in the forms of runoff, subsurface flow and infiltration are intimately related to each other and to precipitation, but it is more complex than is generally realized. The saturated hydraulic conductivity of the soil is the controlling soil parameter. I shall introduce the complexities of the issues with five papers.

Hewlett and Helvey (1970) provided an early review, and some of the first clear data on the effects of deforestation on the storm hydrograph. It will also be useful to readers not familiar with some of the concepts of hydrology to scan the well-illustrated review by Hewlett and Troendle (1975).

Saturated hydraulic conductivity is a highly variable parameter in both space and time. A small quantity of soil fines in runoff water can seriously reduce conductivity. This is throughly reviewed by McDowell-Boyer et al. (1986), a paper which is further enhanced by a "Comment" paper by Germann and Douglas (1987). Retention of small particles within soil pores can reduce permeability by orders of magnitude.

Pilgrim and Huff (1983) found on a large field plot that subsurface storm flow contained over $1 \, g \, l^{-1}$ of fast-moving suspended sediment of $4-8 \, \mu m$ diameter. Their experimental evidence suggested that raindrop impact started the entrainment primarily through soil macropores.

Bonell et al. (1981, 1982) using tritiated water concluded that two flow systems were operating: a rapid by-pass macropore system and a more restricted conductivity in the pore spaces of the clay soils. They were working in a tropical rainforest on a steep slope with a dense mass of exposed roots and rapid litter decomposition. This produced such high rates of saturated hydraulic conductivity that daily rainfalls of 250 mm could be absorbed without surface runoff occurring.

The use of stable isotopes is a promising and incisive advance in partitioning precipitation into its components along the path to a gauging station. Natural variations in isotopic composition of water balance components are used to identify the flow patterns which lead to stream generation in a catchment. Turner et al. (1987) provided an excellent example of this approach as well as references to recent work.

2.3.1 Runoff

The great complexity in evaluating the runoff process lies in the combined variation in precipitation, soil conductivity profile and land morphology. How can this complex be collated or modelled to predict the actual output at the gauging station? Garbrecht and Shen (1988) comment that past research, though partly successful in accounting for morphology, has provided little insight into the physical causes and effects involved. They bravely set out to address this problem, prudently restricting their field to Hortonian drainage networks [having the same bifurcation ratio for all complete subnetworks within it; Scheidegger (1968)]. Their paper will stimulate those who are undaunted by astronomic complexity! For some time we shall have to accept gauging station data at their face value. Unfortunately, gauging stations have more than one face as we shall see below.

The above view is analyzed and substantiated by Freeze (1972a, b). He claimed that the importance of subsurface response of watersheds has been under-rated in most studies up to that time. He explored the mechanism of base-flow generation and the nature of watershed response in base-flow dominated streams through a deterministic model (1972a). For upstream source areas, his modelling (1972b) showed that there are stringent limitations on the occurrence of subsurface storm flow as a quantitatively significant runoff component. Only on convex hill slopes that feed deeply incised channels is subsurface storm flow a feasible mechanism, and then only when saturated soil conductivities are very large. On concave slopes with lower permeabilities, and on all convex slopes, hydrographs are dominated by direct runoff through very short overland flow paths from precipitation on transient near-channel wetlands. On these wetlands, surface saturation occurs from below because of rising water tables that are fed by vertical infiltration rather than by lateral subsurface flow.

Those who are not aware of the complications of runoff will be rewarded by reading the exposition of runoff issues and concepts by Pearce et al. (1986) and Sklash (1990). Storm runoff is *still* a controversial issue (a view supported by the US National Report of IUGG 1979–1982; Wood 1983). In that, and a related paper

(Sklash et al. 1986) they describe the mechanisms involved. They use hydrometric and tracer data for rainfall, soil-water and stream flow and describe the runoff processes and distinguish between "old" and "new" water in the catchment. They are then able to assess the role of macropore flow of "new" water in the system. Their work significantly enchances the understanding of runoff.

Confining their extensive survey of 53 drainage basins to the humid zone, Cordery and Pilgrim (1983) discovered that, contrary to general opinion, storm loss rate is not dependent on catchment size, or soil type or vegetation type. The implications that loss rates vary at random have an important bearing on flood estimation. They also suggest that we do not know what influences the storm loss ratio in the humid zone.

In the tropics Bruijnzeel (1983) adopted a different approach. He studied 40 runoff events on the one small forested basin. Field mappings indicated that contributions to storm flow by channel precipitation, "Horton" overland flow and saturation overland flow originate from well-defined and relatively constant areas in the basin.

The dependence of storm flow on rainfall intensity and vegetal cover was re-examined by Hewlett and Bosch (1984). This is an issue which one would take for granted unless the earlier work of Hewlett et al. (1977) had been studied. They analyzed 1546 storm flows from eight small basins and reconfirmed that hourly rainfall intensity, and consequently overland flow, play a minor role in many source areas. Using regressions, they found that storm-flow volumes showed decreasing sensitivity to rain intensity on high response basins, and as rainstorm size increases on those same basins. That is, small channel-source storm flows respond to hourly intensity but large storm flows from expanded source areas do not. And, of special interest to this review, afforestation of a 195-ha basin with pines produced a barely detectable decrease in storm and peak flows, and no change at all following periodic burning of grass. It follows that one must be wary of plausible assumptions on the runoff process.

In a temperate climate, Burch et al. (1987) describe the difference in runoff between an 80-year grassland and a remnant eucalypt forest. The grassland catchment generated high-peak storm flows and large discharge volumes irrespective of antecedent soil moisture status. The forest gave little runoff provided antecedent soil-water content was below 60% of the available storage capacity. The difference in runoff behaviour of the two catchments was that the subsurface hydraulic conductivities of the grassland soils were about half those of the undisturbed forest at slope positions above the depression areas and even less in the depressions.

De Walle and Lynch (1975) deal with the timing of snowmelt runoff as affected by clearing deciduous forest for harvesting and for reproduction management. Clear-cutting produced significantly earlier and greater daily peak flows from snowmelt. The discharge increment (from time of daily initiation of melt runoff to time of peak flow) was significantly increased. These effects of clear-cutting occurred on both clear and cloudy days.

Having reported the complexities of the runoff process, it would seem unlikely that anyone could develop a simple method of estimating it. Calder and Newson (1979) laid the foundation for a water balance method for annual runoff that could be used for monocultures such as pasture and forest plantations. Pyatt (1984) produced an Information Note based on Calder and Newson's work. He suggested a crude adjustment for deciduous trees, and provided a chart from which to obtain runoff.

2.3.2 Subsurface Flow and Infiltration

When rain commences to fall on soil at a rate faster than its saturated hydraulic conductivity, it ponds on the surface and in so doing air is entrapped. A brief review and some fine experimentation are presented by Constantz et al. (1988). They confirm that air, residing in the transition zone of the soil during ponded infiltration, sharply reduces the imhibition of water. The mechanism is the reduction in the total volume of interconnected pores. Modelling of air entrapment is reported by Kaluarachchi and Parker (1987).

Soil freezing, obviously, also reduces infiltration rate but this can be manipulated because vegetation types can influence soil freezing. Therefore, vegetation can influence infiltration rate. Harris (1972) assessed this effect by measuring infiltration rates under vegetation before and after the soil froze. He used deciduous forest on virgin soil, and both coniferous forest and an abandoned field of grass and forbes on land that had not been cultivated for 6 years. Infiltration was influenced by the number and orientation of connected macropores, and by the extent of blockage of the large drainage channels by ice. The deciduous forest and open field had similar rates. Canopy drip from the closed conifer canopy compacted snow into a solid layer of ice which prevented infiltration.

In south-western Australia, Johnston et al. (1983) exposed the existence of a dual infiltration system in a lateritic subsoil under eucalypt forest in a Mediterranean climate. They infiltrated a dye and traced its course by deep excavation. The dye

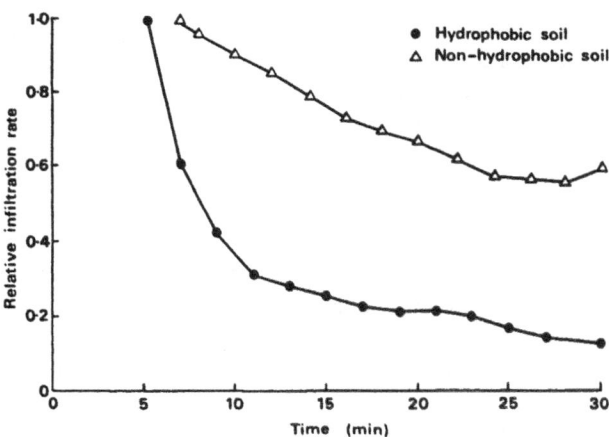

Fig. 4. Time course of relative infiltration rate (infiltration rate/rainfall intensity) under hydrophobic and non-hydrophobic conditions in a forested catchment at Puckapunyal, south-eastern Australia (After Burch et al. 1989)

moved very slowly through the main matrix of kaolinitic clay but rapidly (some three orders of magnitude faster) through perferred channels. Greenwood et al. (1981) also observed a high density of such channels under a pine forest which had been planted after the felling of the prior natural forest. The channels occupied 0.6% of the sample area and ranged in diameter from 1 to 35 mm. Two-thirds of the channels were less than 4 mm in diameter. At a depth of 3 m, all channels contained at least one pine root and as many as eight.

The complexities of measuring, modelling and evaluating infiltration in tropical forests are exposed by the careful work of Wierda et al. (1989). When and where might Hortonian overland flow be predicted to occur in forest landscapes? Weirda et al. recommend that, even though most parameters can be estimated *in principle* from basic soil data, sprinkling infiltrometer field measurements should also be made. That would encompass soil variability due to dynamic surface conditions, macro-porosity, air entrapment, and the irregularity of the wetting front.

I can find no reference to any author reporting uniform infiltration phenomena. My colleagues working on deep "uniform" sands still find variability in wetting. Hydrophobic agents are implicated here. Burch et al. (1989) briefly review the wide range of hydrophobic agents. They present clear effects of the seasonal presence on infiltration and runoff in eucalypt forest and grassland in south-eastern Australia (Fig. 4). How is one to assess readily the presence of hydrophobic agents and to manage them?

2.4 Storage of Soil Water

The rigorous and practicable measurement of seasonal change in storage of soil water became possible with the advent of the neutron moisture meter. It has been used extensively in agriculture for estimating water use by vegetation. It is less popular in forests because access holes need to be much deeper to follow the root system, perennial vegetation requires year-round sampling, the rougher terrain requires more intensive sampling, and decomposing basement rock is more difficult to calibrate.

Neutron probes are appropriate to small, process-study catchments, such as that of McIlroy and Dunin (1982) described in Sect. 2.2.3. They are quite impracticable at basin scale. Kachanoski and deJong (1988) have begun the statistical investigation of scale on soil water storage ranging over a transect 720 m long. Their evaluation of the limits of scale, when published, should be useful.

The neutron probe provides information on soil water status and the rate at which it changes over intervals of 2–3 weeks. It does not permit one to discover how that status was achieved. The use of stable isotopes of hydrogen and oxygen can greatly assist in studying the partitioning of soil water into infiltration, eva-poration, recharge and mixing. That capability is reviewed in a significant paper by Barnes and Allison (1988) – an effective and fruitful mixture of statement and of application by modelling with several applications. Experimental results of the influence of deforestation and revegetation on soil water storage are presented in Sect. 4.4.

2.5 Recharge, Leakage and Barriers

This is the province of the hydrogeologist and brings us to a wider, interpretive perspective of water balance. The flow of water into and out of an aquifer, the flow of water between aquifers, and geologic barrages which divert the flow of water into an unexpected direction, are important issues on the larger scale. Soil water balances traditionally go no further than recharge to and discharge from perched aquifers and the unconfined permanent aquifer (often called the phreatic aquifer which is the one containing the water table). And it is usual to assume, tacitly, that an aquifer identified within the topographic boundaries of a catchment will discharge at its outlet. If it were not for developments in geophysics (e.g. seismic, electro-magnetic induction, transient electromagnetic induction), we might not know what complications exist below ground. That is, unless a sufficiently dense bore field allowed the accurate assessment of potentiometric heads from which flow directions can be inferred.

2.5.1 Recharge and Discharge of Aquifers

Aquifers which possess a water table (phreatic aquifers) will be recharged from infiltrating soil water and that rate can be measured. In parts of the catchment the phreatic surface may be confined by impervious, or slightly pervious, material which will limit the infiltrating soil water and therefore the rate of recharge to the aquifer. This must be accounted for, so it is important to identify confining strata. The phreatic aquifer may leak into a lower aquifer or vice versa. Should these phenomena occur and are not accounted for, then the estimates of aquifer performances will produce errors in water balance.

Increased recharge after forest clearing shows as a rise in water table, a rise which continues over several years until a new equilibrium (water balance) is attained (e.g. Peck and Williamson 1987; Sect. 4.5).

Almost every aspect of groundwater recharge is covered in a symposium edited by Sharma (1989) who also presents a paper on methods of measuring recharge by the use of natural tracers (Sharma and Craig 1989; Sect. 4.5).

Table 2. Comparison of recharge values estimated by the injected tritium method with values estimated by hydrogeologic parameters (After Athavale et al. 1980)

Site	Recharge (percent rainfall)	
	Injected tritium method	Hydrological method
Warangal	8.3	8.5
Kamalapur	8.7	9.4
Parkal	6.1	8.4
Bhopalpally	15.5	23.0

A technique for assessing recharge of phreatic aquifers is the tritium injection method. For those unfamiliar with tritium, Athavale et al. (1980) give a good exposition and sets of data from India. Their data had high reproducibility (2–10%) and they correlated well with the more costly and cumbersome well network method (Table 2).

It is obvious that recharge areas (above phreatic aquifers) need to be mapped. Stoertz and Bradbury (1989) have assessed a method of mapping recharge using a groundwater flow model based on the USGS Modular Groundwater Flow Model of McDonald and Harbaugh (1984). The mapping compared well with conventional field observations but the actual magnitudes of recharge rates were less reliable. They overcame that problem by introducing stream discharge data into the model which produced net recharge rates averaged over the basin. This seems a sensible approach provided there are no leaks!

2.5.2 Leakage of Aquifers

While water balance hydrologists may ignore catchment leakage, hydrogeologists expect it. Anyone who has attended a mine water conference becomes actuely aware of that phenomenon.

Langford et al. (1980) inferred leakage in 3 out of 14 catchments, based on the magnitudes of error terms in a classic analysis of catchment covariance. Bren and Leitch (1986) also used the statistical approach on three neighbouring catchments. One of them yielded more than the other two. How is this to be interpreted? How much is the difference due to leakage and how much to capture?

2.5.3 Barriers to Flow

A gentle landscape may belie a startling complexity beneath the surface. Engel et al. (1987) provide an example which should convince water balance researchers of the value of such information. Barriers such as intruding dykes are precisely delineated if they have a slight magnetism (Fig. 5). Aerial photographs of bare ground are also revealing. A surrogate parameter in the form of salinity contours delineates recharge areas. The current development of airborne instruments developed by Buselli et al. (1986) will provide a practicable and cost-effective aid to water balance in the future – an optimistic note on which to close this critique of water balance methods.

2.6 Modelling Water Balance

Modelling is crucial to water balance research. There is no other option, bearing in mind:

* the high variability within and between catchments;
* the large areas, volumes and periods involved;
* the array and complexity of influential and interacting parameters.

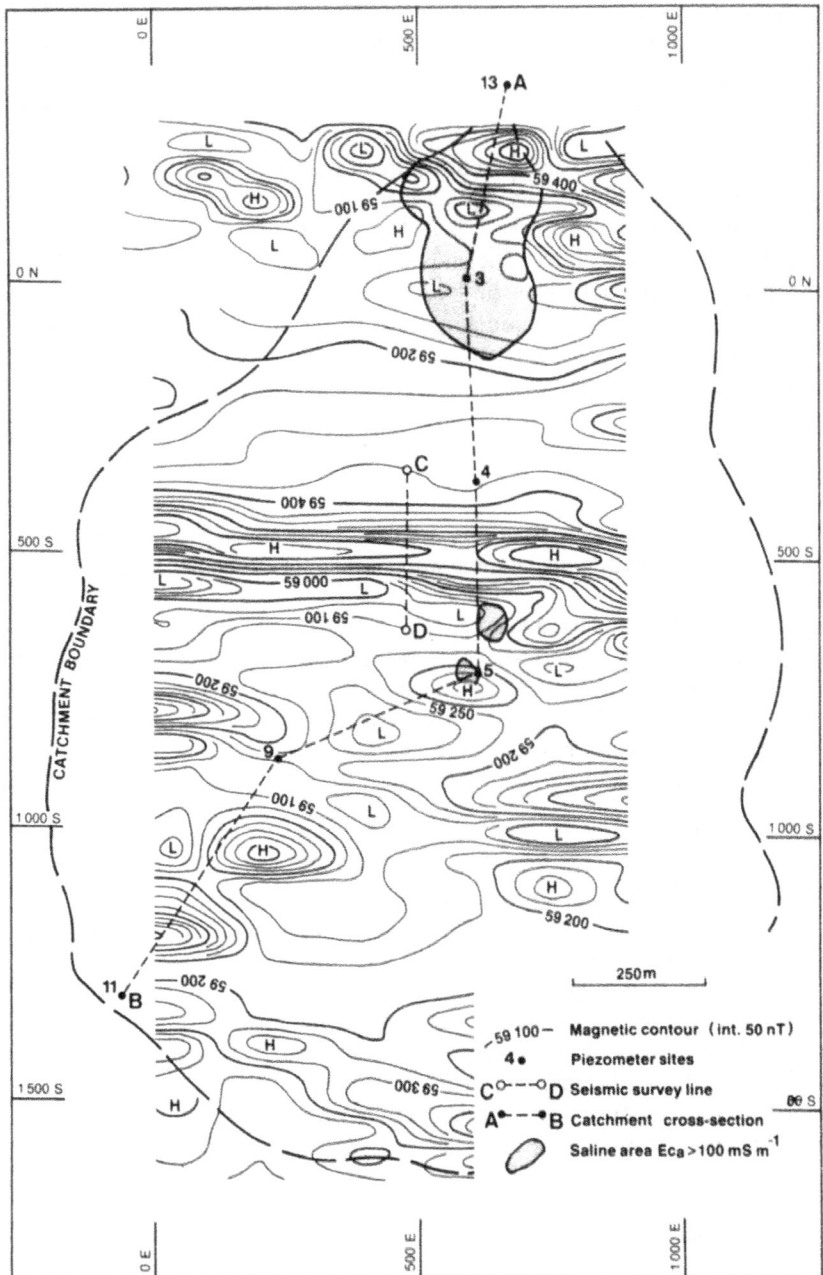

Fig. 5. Below-ground heterogeneity as revealed by magnetic, soil resistivity and electrical conductivity surveys (After Engel et al. 1987)

Two daunting issues confront each other. The first is the provision of data and information to serve the model. The second is to devise modelling processes which adequately represent the interacting phenomena as perceived by the field scientist.

The most enthusiastic optimist among us must admit that, in the field of water balance, our overall record has been poor, i.e. inadequate measurement and sampling, leading to, at best, plausible concepts of interacting processes which yield an intractable data base. The modeller is forced to compromise, for which the modeller is blamed rather than the provider of the data. This is not a moral imputation; it is simply bad research management, which arose from the immensity of the task and the rodent-like manner with which we used to set about it.

The history of water balance research started with individual, or small groups of, researchers developing field techniques. They were applied on conveniently small scales and on sites which were not always representative, and reported as isolated sets of data in journals. Though techniques improved, the type of information derived was inadequate to be applied realistically to all but the smallest and simplest catchment. Larger research teams assembled eventually, with biologists sometimes being segregated from physical scientists; modellers were called in at a late stage as separate contributors and were received without deference unless it was the measurer who became the modeller.

Now, I believe, we manage water balance projects better. We are encouraged to form multidiscipline groups and we have some broadly based journals.

The abstract of Bevan's (1989) critique "Changing ideas in hydrology – the case of physically based models" deserves to be quoted (and the paper read) in full.

"This paper argues that there are fundamental problems in the application of physically-based models for practical prediction in hydrology. These problems result from limitations of the model equations relative to a heterogeneous reality; the lack of a theory of subgrid scale integration; practical constraints on solution methodologies; and problems of dimensionality in parameter calibration. It is suggested that most current applications of physically-based models use them as lumped conceptual models at the grid scale. Recent papers on physically-based models have misunderstood and misrepresented these limitations. There are practical hydrological problems requiring physically-based predictions, and there will continue to be a need for physically-based models but ideas about their capabilities must change so that future applications attempt to obtain realistic estimates of the uncertainty associated with their predictions, particularly in the case of evaluating future scenarios of the effects of management strategies."

Hetrick et al. (1986) compared the ability of the relatively simple soil compartment model – SESOIL – with that of the more data-intensive terrestrial ecosystem hydrology model – AGTEHM – to predict seasonal and annual values of infiltration, total evaporation, surface runoff and groundwater runoff. The comparisons were performed using data from a wide range of vegetation systems: a deciduous forest watershed, a grassland watershed and two agricultural field plots. Good agreement was obtained, for simple reasons which are explained, between the predictions by the two models except for moisture content. In addition, the SESOIL model was used to predict the measured values of evaporation and runoff. This it did well for the annual values but was sometimes inaccurate on a seasonal basis. They concluded that SESOIL is a valuable screening model but added the warning that, although the predictions of the two models agree well, it should be noted that neither model has been extensively verified with actual data not used during calibration.

Alley (1984) investigated the use of regional water balance models to transform monthly precipitation and monthly 'potential evapotranspiration' data to monthly and annual runoff estimates. He used 50-year records of monthly stream flow at ten stations. Three two- to six-parameter models were examined, variants of the Thornthwaite–Mather model, the Palmer model and the Thomas *abcd* model. Alley explored four important issues in evaluating these models:

* Are the identified parameter values reasonable and consistent between models?
* Do any of the models result in substantially lower prediction errors than the others, and are calibration errors a reliable indicator of prediction errors?
* How accurately can the parameter values be estimated and what are their covariances?
* What is the relationship of modelled state variables which are related to ground-water storage, to actual measurements of water levels in wells?

Prediction errors are similar between the models. Simulated values of state variables, such as soil water, differ substantially among the models and fitted parameter values for different models sometimes indicate an entirely different type of basin response to precipitation. Problems in parameter identification arise, including:

* Difficulties in identifying an appropriate time lag factor for the Thornthwaite–Mather type model for basins with little groundwater storage;
* Very high correlations between upper and lower storages in the Palmer-type model; and
* Large sensitivity of parameter *a* of the *abcd* model to bias in estimates of precipitation and "potential evapotranspiration".

Alley considered that these and other results suggest that extreme caution should be used in attaching physical significance to model parameters and in using the state variables of the models in indices of drought and basin productivity.

3 Influence of Site Operations on Water Balance

The mechanical nature of logging and revegetation operations strongly influences water balance and sediment transport. Accordingly, I want to separate this issue from the direct effects of vegetation change per se (Sect. 4).

The review paper by Greacen and Sands (1980) covers the mechanics of soil compaction, the effects of compaction on physical properties of soil, the consequences for root growth and techniques for prevention and amelioration of compaction. They provide quantitative data on contact pressures of a range of logging machines, isograms of vertical stress, and a table of contact pressures for each type of operation and numbers of passes (in Australian forests). All other authors restrict themselves to a qualitative evaluation of processes.

The world wide examples of site operations which follow are grouped on a climatic basis. Where the climate has not been specified, I have consulted the Times Climatic Atlas.

Bruijnzeel (1986) gives a wide and critical view of major forest uses and conversions in the humid tropics: shifting agriculture; commercial logging; conversion to forest tree plantations and extractive tree crops; conversion to grassland; conversion to annual cropping and agroforestry; reforestation.

Bruijnzeel tabulates published data on forest cover transformation and changes in water yield for the humid tropics, and concludes that:

1. The general finding of Bosch and Hewlett (1982) that deforestation increases water yield (up to 450 mm yr^{-1}) applies;
2. Regardless of type conversion, the highest increases are usually in the first year followed by a decline with the advent of new cover;
3. The new equilibrium water yield is higher with grasslands, tea plantations and moderately stocked conifers in Indonesia; is the same with dense pines in East Africa after 9 years; is lower with eucalypts in Madagascar.

After reviewing the processes involved, Bruijnzeel concludes that deterioration of land resources after deforestation in the humid tropics is not so much the result of deforestation per se, but rather to *poor land-use practices afterwards*. Forest conversions need not lead to land degradation provided logging roads and culverts are properly planned, riparian buffer zones are established, steep slopes are terraced and water ways are controlled.

Long-term (1958–1974) catchment experiments in East Africa (Table 3), ranging from tropical humid savannah to warm humid, marine, west coast climates are reported by Edwards and Blackie (1981). These were rigorous experiments which yielded the following conclusions for this section:

1. Replacement of rainforest by tea plantations produced no significant runoff or sediment loss due to efficient soil conservation measures.

Table 3. Effects of long-term changes in vegetation in East Africa on total annual evaporation, AE, expressed as a proportion of pan evaporation, EO (After Edwards and Blackie 1981)

Location	Catchment	Dominant vegetation	Period	Mean rainfall (mm)	SEE[a]	AE/EO	SEE[a]
Kericho	Lagan	Montane rainforest		2219	± 149	0.93	± 0.032
	Sambret subcatchment	Bamboo	1967–73	2026	± 140	0.86	± 0.022
	Sambret	Tea		2011	± 139	0.84	± 0.030
Kimakia	C	Bamboo		2143	± 158	0.76	± 0.012
	A	Pines	1967–73	1997	± 151	0.76	± 0.020
	M	33% Grass		2062	± 137	0.70[b]	± 0.023
Mbeya	C	Montane rainforest		1924	± 143	0.93	± 0.065
	A	Cultivated crops	1958–68	1658	± 120	0.64	± 0.025

[a] Standard error of estimate.
[b] R-Q only.

2. Replacement of bamboo forest by pine plantation initially decreased water use (warm humid, marine, west coast climate). Once the pine canopy closed, there were no significant differences in water use and sediment yield. The effect of pine felling remained to be observed.
3. Replacement of evergreen forest by small-holder cultivation on very steep slopes greatly increased water yield (tropical, humid savannah). Lack of runoff and sediment yield was attributed to porosity of the ash-derived soils.
4. Bush clearing followed by gross colonization and exclusion of cattle and subsequent controlled grazing increased infiltration and reduced peak flows (tropical, humid savannah to warm humid, marine, west coast climate).

The foregoing results indicate that, with careful management, quite drastic changes in vegetation can be accomplished without site degradation in the tropics of Africa.

Lal (1981) examined the effects of different clearing and post-clearing methods on sedimentation, runoff, soil erosion, and the grain yield of maize in tropical Nigeria (Table 4). These values speak for themselves but do not, Lal reminds us, expose the great inefficiency of manual methods.

The more specific effects of clearing treatment on soil in similar circumstances are provided by Lal and Cummings (1979). Compared with manual clearing, the mechanical methods increased bulk density and penetration resistance and reduced infiltration and saturated hydraulic conductivity.

In a humid, subtropical climate (1070 mm evenly throughout the year), Blackburn et al. (1986) used nine small forested (*Pinus echinata* and mixed hardwoods) watersheds in East Texas to determine the effect of harvesting and site preparation on storm flow and sediment loss (Fig. 6). Three replications of three treatments were used: (1) clear-cutting, followed by shearing, windrowing, and burning; (2) clear-cutting, followed by roller chopping and burning, and (3) undisturbed control. The treatments induced very large differences in storm flow and sedimentation.

Bren and Leitch (1985) reported the effect of roads in a forest in south-east Australia (constantly moist, marine, west coast climate). Runoff from a stretch of forest road was measured continuously both before and after passage across a section of undisturbed forest. The volume of storm runoff per unit road area was found to

Table 4. Effects of methods of deforestation and tillage systems on sediment density, water runoff, and soil erosion from maize-cassava rotation in a tropical rainforest in south–west Nigeria (After Lal 1981)

Clearing treatment	Tillage system	Sediment density (gl^{-1})	Water runoff $(mm\,yr^{-1})$	Soil erosion $(t\,ha^{-1}\,yr^{-1})$
Traditional clearing	Traditional seeding	0.0	2.6	0.01
Manual clearing	No tillage	3.4	15.5	0.4
Manual clearing	Conventional tillage	8.6	54.3	4.6
Crawler tractor/shear blade	No tillage	5.7	85.7	3.8
Crawler tractor/tree-pusher	No tillage	5.6	153.1	15.4
Crawler tractor/tree-pusher	Conventional tillage	13.0	250.3	19.6

Sediment density reported here was from a rainstorm monitored on 31 May 1979.

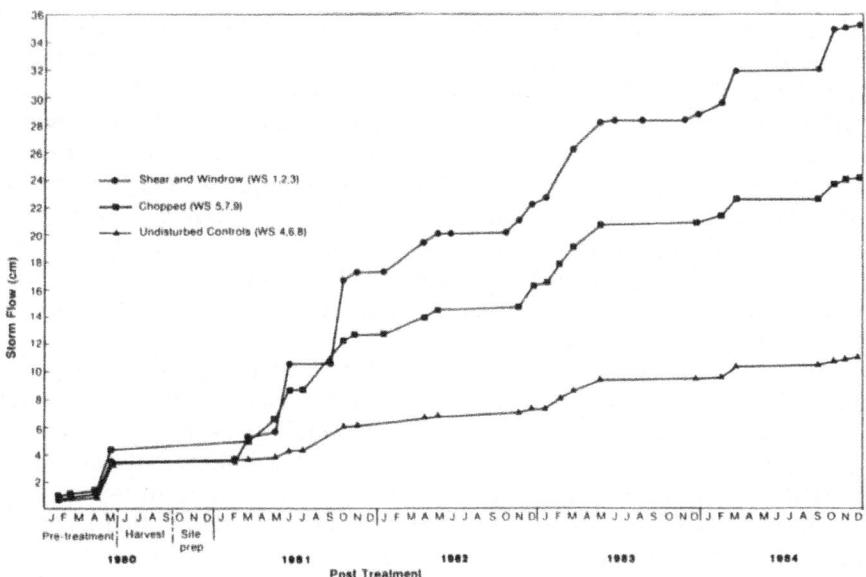

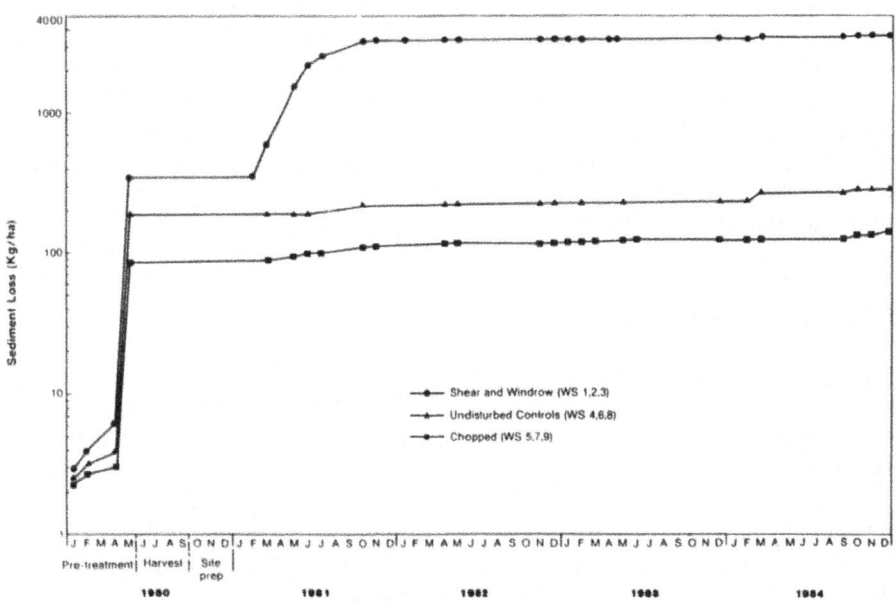

Fig. 6. Cumulated storm flow and sediment loss by treatments for the pre-treatment year and four post-treatment years, Alto, Texas, USA (After Blackburn et al. 1986)

be predicted best by the depth of rainfall. The peak flow per unit road area was predicted best by the maximum 60-min rainfall intensity. The results were compared with models of storm-flow and peak-flow generation derived from flow data for a small forested catchment near the road stretch. It was concluded that the presence of a length of road in this catchment would lead to more storm flow for small and moderate storms but would make little difference for larger storms. However, the road would possibly make a substantial contribution to the peak flow for all storm sizes, although timing differences in reaching peaks could complicate this.

Harr et al. (1975) found in the forest of the Oregon Coast Range, USA, that if roads occupied more than 12% of the catchment, peak flows, but not volumes, of storm hydrographs increased significantly.

In those parts of northern Europe and Britain with a climate ranging from marine, west coast constantly moist through cool humid to subarctic constantly moist, land drainage is an important site issue (Robinson 1989). It may be necessary to drain land before planting forests. Thus, the hydrologic effects of the developing forest are confounded with the drainage effect with time. That is, the trees improve hydrologically with age, whereas the ditch system will deteriorate, and undrained may perform differently to drained plantations.

Luval et al. (1985) applied the ultimate research-oriented treatment to a forest classified as being a transition between tropical wet and premontane wet forest ($4000 \, \text{mm} \, \text{yr}^{-1}$). A quarter hectare was hand-felled using machetes and chain saws only. All material was left on the plot, thus minimizing soil disturbance. The plot was allowed to regenerate.

Daily total evaporation was measured on the control and the plot 6 months before and after the treatment, by which time total evaporation had already reached 80% of the control. The authors concede that their treatment had little applied relevance. What would Shuttleworth's (1988b; Sect. 2.2.3) model have predicted for total evaporation, had the cleared plot been say 100 km^2?

4 Effects of Deforestation and Revegetation on Water Balance

Generally, the measurement of water balance parameters is either inaccurate or impracticable on a large scale. That need not hinder a useful assessment of the effects of vegetation treatment on the water cycle. The magnitude and depth of reported findings is encouraging.

The greatest interest has been shown in the humid tropics. We are fortunate in the timely publication of Bruijnzeel's (1990) comprehensive and lucid review and synthesis "Hydrology of Moist Tropical Forests: A State of Knowledge Review". It updates and extends this brief 1986 review which is used in this section.

4.1 Precipitation

4.1.1 Rain

In his review of environmental impacts of (de)forestation in the humid tropics, Bruijnzeel (1986) assesses, quantitatively, the importance of forests in the recycling of rain and the consequences of conversion to agriculture. Two interrelated questions arise:

1. Does a humid forest induce more rain to fall than would occur if some other form of vegetation, say agricultural, were to be established?
2. Is the water, which is evaporated from wet leaves (interception), from transpiration and from the soil surface, recycled under the high energy regime of the tropics to produce more rain either at that locality or elsewhere?

Bruijnzeel concluded that only in Amazonia and in 'cloud forests" is recycling likely to be important. The Amazonian evidence was provided by Salati et al. (1979), who used the stable isotope composition of rain to resolve the issue. Although they found the Amazon basin to be hydrometeorologically more complex than expected, the small inland gradient of the isotope composition of the rain established the magnitude of reevaporation.

Salati and Vose (1984) describe the water cycle, nutrient balance, erosion, heat balance and climatic change and the potential consequences of disequilibrium in the Amazon Basin. Lettau et al. (1979), dividing the region into 5° longitudinal segments, concluded that a substantial proportion of rain in the west falls at least a second time. Brooks (1985) estimated that even if 40% of the forest was cleared for agriculture it would reduce rainfall by only 6%. On that evidence, Bruijnzeel concluded that the effect of clearing Amazonian forest on rainfall would not be serious. It would be sensible though to establish a vegetation system which maximizes recycling.

Bruijnzeel (1986) was more definite when reviewing "cloud forests", concluding that their replacement by agriculture would be disastrous. It would be advisable then to replace the forest with a vegetation system which allowed some cloud and fog "stripping" to occur. Bruijnzeel concluded that it is unwise to expect firm conclusions on the possible climatic impacts of forest removal in the tropics. More definite work is needed on the parameterization of land surface processes in climate models, which he proposed to be done by meteorologists and hydrologists working together. That also is the attitude of Shuttleworth (1988b) in the 2-year Anglo-Brazilian study of micrometeorology and process hydrology in Amazonia. In that project (see also Lloyd and Marques 1988; Lloyd et al. 1988), approximately 10% of rainfall was intercepted by the forest canopy, and this accounted for 20–25% of the evaporation. The remainder occurred as transpiration from the trees. Over the same period, about half of the incoming precipitation is returned to the atmosphere as evaporation, a process which requires 90% of the radiant energy input. These proportions vary seasonally in response to the large variation in rainfall. The average evaporation over 2 years was within 5% of potential evaporation. Monthly average evaporation exceeds potential estimates by about 10% during wet months, and falls below such estimates by at least this proportion in dry months. These data provide the hitherto missing numerical basis for assessing the likely consequences of Amazonian deforestation on surface water and energy balances.

In the absence of a significant change in climate and on the basis of these data, a land-use change to stable grassland (or bare soil) is likely to reduce Amazonian annual average evaporation by 10–20%.

Leopoldo et al. (1987) again considers the hypotheses of recycling of rain, its importance for water balance and the possible import from and export to regions and oceans, as against continuous recycling where it arises in the forest. They return to the isotope data of Salati et al. (1979) above. The proportion of rain coming from transpiration (evaporation?) is very high and the period of the cycle is about 5.5 days.

Apart from Shuttleworth's (1989b) estimate of recycling of rain from grassland and bare soil, we still lack the magnitude of recycling by types of candidate vegetation to replace the forest in Amazonia.

I can find no references in the literature to the above issues in other climatic regions, yet such investigation is urgently needed.

4.1.2 Snow and Fog

The paper by Golding and Swanson (1986) is a good introduction to the influence of reforestation on snow hydrology. Snow falling on forest canopies tends to be evaporated (interception) and lost to the catchment. Snow in clearings tends to melt. Further, some of the snow in clearings has arrived by wind from surrounding forest. There are other complications – sun aspect and exposure – which were evaluated in the Rocky Mountains in Alberta, Canada. Their objective was to manage snowmelt by prolonging recession flow and to delay the time of peak runoff by prolonging snowmelt.

Golding and Swanson found that snow accumulation was 13–45% greater in clearings (which ranged from 12 ha to a few hundred square metres) than in the forest. There was no evidence of redistribution of snow from the forest to the large clearing on one catchment, so increased melt is attributed to the elimination of interception. On another catchment, increased yield of water seemed to be due entirely to redistribution. I do not think that the authors would claim that they had as yet completely achieved their snow management objectives in the above exercise.

In western Oregon, USA, annual rainfall is about 2300 mm, 80% of it falling between October and April during long-duration, low intensity frontal storms. At elevations of 400–700 m, snow is common but rarely persists beyond 2 weeks and generally melts in 1 to 2 days. Harr and McCorison (1979) concluded that logging altered snow accumulation and melt sufficiently to delay and to decrease peak flows. They attributed this to snow persisting longer in the cleared area. They observed that snow in tree crowns had a greater surface area exposed to moist, moving air than did snow on the ground in cleared areas. Annual peak flow was reduced by 32%, and average delay of all peak flows was nearly 9 h due to clear-felling.

Calder (1986a) and his colleagues measured and modelled the changes in water balance in upland (> 500 m) catchments in Britain arising from large-scale afforestation. His analysis is that snow on grassland or moorland, being relatively smooth, inhibits turbulent transport of heat and vapour. This, together with a high albedo, inhibits evaporation to less than 0.06 mm h^{-1}. These low rates are counterbalanced by daily periods of condensation. About half of the energy for the spring snowmelt was supplied by sensible heat transport during storms and the remainder by radiant energy.

Corresponding values for snow on forest canopies were derived from three types of equipment: (1) gamma-ray attenuation to measure change in density, whence change in water content of the canopy; (2) weighing equipment to measure canopy storage of a single tree; (3) a heated plastic sheet to measure net precipitation from snow or water from the canopy. The results of this work confirm the observation of Harr and McCorison above on snow interception from forest and demonstrate that evaporation losses from trees in snow may match or exceed loss from wet canopies, and that the reverse holds for short, "smooth" vegetation.

The review paper of Morris (1989) is of critical relevance to this section. It deals with issues of sublimation of blowing snow and the application of turbulent transfer theory on the catchment scale. But it is not specific on vegetation types, so I simply draw your attention to it.

Extensive, early studies on the effect of vegetation on snow in the subalpine regions of south-east Australia were reviewed by Costin (1967). The snow water content under different vegetation communities were: snow gum (eucalypt), 1100 mm; dead snow gum, 530 mm; groundwater communities, 810 mm; herbaceous communities, 960 mm. Annual rates of deposition of rime on naturally timbered areas ranged from $50-125$ mm yr^{-1}. This was considered a net gain to the catchment.

4.2 Evaporation

A major issue governing change in evaporation with change in vegetation is the latter's structure. I refer to the presence of one, two or three storeys–ground flora, mid-canopy and upper canopy species. Although all three types might produce notionally the same annual total evaporation rate, after logging, the single-storey forest has no residual vegetation; the double-storey forest still has its ground flora; the triple storey may have even more residual vegetation. Accordingly, the three forest types would have very different post-logging rates of evaporation unless there was immediate clearing, firing and conversion to agriculture or silviculture. The following results surprised me.

Starting with Mediterranean climates of south-western Australia, the jarrah (*Eucalyptus marginata*) forest has three storeys, but it is sufficiently open for direct sunlight to reach the ground flora in many small areas, which of course move with the time of day. The rainfall is 1000–1200 mm at its western boundary. The water table is at about 20 m but the forest is probably phreatophytic. Greenwood et al. (1985b) measured total evaporation of the lower and middle storeys with ventilated chambers and found that together they discharged about 50% of the annual rainfall. We noted that whenever direct sunlight reached a ground flora sampling site, there was an immediate surge in evaporation. It was of such magnitude that had the entire upper storey been logged, the understorey could have compensated substantially for the trees. Such situations occur with forest roadside clearing and powerline transects, but not clearing for argiculture, industry or urbanization.

At a more inland and drier (800 mm yr^{-1}) region of the jarrah forest, large-scale clearing for agriculture has occurred, the water table is shallower (5–10 m) than the previous example, and the subsoil has accumulated cyclic salt above the water table. The annual pastures and crops grow only over the wet 6 months (May–October), so the discharge by evaporation is much less than from the prior forest. The consequence

Table 5. Annual rates of total evaporation from grazed pasture and *Eucalyptus* species in plantations in south-western Australia. Rainfall for the year was 680 mm (After Greenwood et al. 1985a)

Site	Vegetation	$E(\text{mm yr}^{-1})$
Upslope	Pasture	370 ± 50
	E. cladocalyx	2660 ± 140
	E. globulus	2690 ± 40
	E. maculata	2330 ± 320
Midslope	Pasture	410 ± 30
	E. globulus	2210 ± 360
	E. leucoxylon	1840 ± 390
	E. wandoo	1620 ± 190

is extra recharge to the water table which then rises into a saline zone. The mobilized salt leads to land and stream salinity on a large scale.

On 12% of a catchment in such farm land, Biddiscombe et al. (1981) estblished highly advective tree plantations, mainly eucalypts, to discharge more groundwater via transpiration, and to reduce recharge by increasing interception. On these plantations, Greenwood et al. (1985a) measured total evaporation from crops, pastures and the more promising species of eucalypt when they were 8 years old (Table 5). The crops and pastures discharged about 400 mm yr^{-1} of the rainfall, whereas the best eucalypts evaporated about 2500 mm yr^{-1}. This sixfold difference was ascribed to the deep phreatophytic root system of the eucalypts and to greater leaf area duration (ever green) and advective energy. That advective energy expressed itself in the measurement of rainfall interception which averaged around 25% of rainfall for the trees. Interception is high in Mediterranean climatic zones because the winter rain tends to be associated with frontal systems which induce rapid wetting and drying cycles.

Greenwood et al. (1981) also measured total evaporation from an agroforestry system with high-pruned *Pinus radiata* (16 years old and 16 m high) and *Trifolium subterraneum* pasture grazed by sheep, again in south-western Australia. The land had formerly carried virgin jarrah forest. The annual evaporation equalled the rainfall of 900 mm yr^{-1}. No evidence of a perched water table was found. Measurements of soil water with a neutron meter showed that no water was extracted below 4.5 m. The results imply that the agroforestry system had restored the cleared land to a water balance similar to the prior forest with possibly a greater evaporation component than before.

In the same climate, but in a quite different ecosystem, Farrington et al. (1989) measured annual evaporation from a low natural woodland on a coastal sand plain. Beneath this plain was a high-quality groundwater mound supplying Perth city. The woodland had three storeys – ground flora, shrubs and small trees (*Banksia* spp.) – which, together with the soil, evaporated 666 mm from the annual rainfall of 863 mm. Total evaporation was 427 mm from the ground flora, 97 mm from the shrubs and 142 mm from the trees. Such high evaporation from the ground flora was unexpected. Who can claim to be an accurate assessor of evaporation from components of mixed vegetation without measurement? Yet, it is often requested by hydrology consultants!

Some of the foregoing examples eatablish that it is possible to change vegetation without affecting the water balance except perhaps in the short term. I place them first because evaporation was measured directly and not estimated by inference. The alternative is to measure the other components of water balance from which total evaporation can then be inferred, as discussed in Sect. 2.2.3. My colleagues, Sharma (1984) and Williamson et al. (1987) took that classic approach in the jarrah forest of the Collie River basin, south-westren Australia.

Sharma's work was located in the $1100 \, \text{mm yr}^{-1}$ rainfall region. For each of the 5 years he tabulated water balance parameters on a 6-monthly basis – October to March (hot, dry) and April to September (cool, wet) – the parameters were precipitation (substantially rain), stream flow, change in soil water storage, change in groundwater storage and leakage either into or out of the catchment. Leakage was considered negligible as concluded from groundwater contours (Sharma et al. 1982). Averaged over the 5 years the rate of total evaportion was $954 \, \text{mm yr}^{-1}$ or 93% of rainfall. He considered this to be a high rate and due to strong interception from the hilly catchment and to high rates of transpiration.

Concurrently, total evaporation from a similar forested catchment was measured for 3 years. Then it was cleared for agriculture. Annual evaporation derived from the water balance equation averaged 89% of rainfall for the forested years, a value which fell to 63% on average over the 7 agricultural years (Williamson et al. 1987).

Williamson et al. also present results for a similar, concurrent water balance exercise involving two partially cleared catchments in the $800 \, \text{mm yr}^{-1}$ zone of the jarrah forest. Again, a 3-year calibration period before, in this case, partial clearing, and again the measurement of interception and an estimate of transpiration from the water balance residual were made. From those data, I have computed that total evaporation from both catchments before and after clearing was 97%. One of those clearing strategies was a combination of parkland and strip clearing; the other was complete clearing of one-half of the catchment. As the catchments were small ($1–4 \, \text{km}^2$) the clearing may have induced advective energy. Another explanation may be the much lower rainfall of the area (730 mm compared to 1030 mm in the previous example). If most of the rainfall can be accounted for by total evaporation, there cannot be much left to partition between the other water balance components: soil water, runoff, and recharge. On the other hand, it is reassuring that the evaporation term is so "resilient" between vegetation systems.

Two papers from subtropical eastern Australia ($1100 \, \text{mm yr}^{-1}$) describe the change in evaporation when pasture is converted to agroforestry, the reverse of the previous example. Eastham and Rose (1988) used lysimeters to measure evaporation from pasture under increasing tree density. As tree density increased, so the root-length density of pasture became significantly lower. This was reflected in low values of evaporation from pasture. On the other hand, the denser canopies protected the grass from winter frost.

The second paper (Eastham et al. 1988) describes water balance with respect to the trees, *Eucalyptus grandis*, and the soil. As tree density increased, so evaporation increased and drainage losses declined. At the closest spacing, soil water, as measured by the neutron probe, reduced the wilting point down to 5.6 m. At the wider spacings, individual trees were able to exploit a large volume of soil water stored in the surface layers so that soil water depletion at depth was less. A water balance table is included.

Kelliher et al. (1986) used a two-layer canopy model to estimate the effect of understorey removal from a Douglas fir forest. Calculated values of evaporation agreed to within $0.2\,mm\,day^{-1}$ of values using the highly regarded Bowen ratio-energy balance technique. The removal increased tree transpiration by $0.4\,mm\,day^{-1}$. The model used was Shuttleworth's (1979) development of the Penman–Monteith evaporation equation for multilayer, partially wet forest canopies, modified to account for having stomates only on the underside of their canopy.

Parker et al. (1985) reported on the same project as Luval et al. (1985; Sect. 3), but dealt with the hydrologic effects of cleared, regenerating tropical rainforest in Costa Rica (Table 6). The forest had been cleared by hand with extraction of timber or burning. Total evaporation was monitored for 13 months after clearing. By that time, the cumulative total evaporation in the regenerating forest was 70% of that for the mature forest. Occasionally, total evaporation rates approached those of the control during the early months after felling. On the other hand, the components of evaporation changed over the period. Starting with 100% contribution from soil evaporation during the first month, this component declined with the rapid development of regrowth from stumps. Transpiration then became the dominant component, followed by interception which of course increased with leaf area; in periods of exceptionally high rainfall, interception exceeded transpiration.

Table 6. Summary of the hydrologic budgets for a mature rainforest and $2500\,m^2$ clear-cut on hillslopes in north-eastern Costa Rica. Tabled values are the percentages of precipitation in that hydrologic compartment. Precipitation for the 404-day period following forest felling (March 1983–May 1984) was 3673.9 mm (After Parker et al. 1985)

Hydrological compartment	Mature forest	$2500\,m^2$ Clearing
Throughfall	86.8	88.5
Stem flow	3.8	0.7
Canopy interception	9.4	10.7
Litter interception	0.9	1.0
Total interception	10.3	11.7
Net precipitation	89.7	88.3
Soil evaporation	1.0	1.4
Transpiration	54.5	32.8
Evapotranspiration	64.8	44.5
Total evaporation	65.8	45.9
Change in soil storage	− 2.2	− 1.0
Percolation	36.4	55.2

The examples above demonstrate the sensitivity of total evaporation to changes in vegetation. Treatments can be devised to reduce, increase or equal prior evaporation. It is not generally realized that such effective management options are available. It follows that the effects of deforestation on water balance may be offset by an ensuing revegetation program.

4.3 Surface and Near-Surface Fluxes

Runoff, subsurface flow, infiltration, stream flow and water yield are integrated parameters. They should be considered together were it not for the numerous references before us and that specific processes may fare differently with vegetation regimes. Another problem in reviewing arises from the dependence of the hydrologic outcomes on the detailed characteristics of the catchment which are not feasible to present. Therefore, I must write a somewhat anecdotal, rather than a deterministic, account of crucially important issues. At best, it provides an inventory of climate, circumstance, treatment and result, from which to collate a reading list of your interests. And we must always keep in mind that every catchment reported has secrets carefully sequestered by the rigours of authorship!

4.3.1 Runoff, Subsurface Flow and Infiltration

It would be of immense value if a catchment manager could have at his disposal a practical model which would predict the outcome of vegetation changes on water fluxes. By practical, I mean readily obtainable data sets from the catchment to be modified. This is what the Institute of Hydrology, Britain, set out to achieve through a lumped conceptual model using estimates of daily rainfall, flows and total evaporation (Gross et al. 1989). From work in Britain and France, they conclude that even with limited data, which is almost universally the most we can expect to be available, coarse predictions of the effects of vegetation treatment on flow regimes can be estimated. By applying low flow analysis techniques to the flow data produced by models of different proportions of forest in a catchment, more details can be obtained. The combination of the lumped conceptual model and low flow analysis techniques would be valuable in any preliminary analysis of the effect of vegetation change, and provides an encouraging start to this section. Their colleagues, Blackie and Newson (1986), in retrospect, reinforce and extend their findings, whild Blackie (1987) presents some basic data and makes the important (though often neglected) point that water authorities have accepted their work.

To the African tropics first with Mumeka (1986), who presents the effects of deforestation, followed by subsistence agriculture on runoff on the headwaters of the Kafue River, Zambia, over several years. The climate has pronounced wet (November to April) and dry seasons. Rainfall is 1400 mm yr^{-1} and altitude is 1300 m. During the first phase (1966–1974), several catchments were left in their virgin-forest state for matching. In 1974, two catachments were cleared in the dry season. Five families were settled on each on them. Each family was allotted 2 ha on which to grow, traditionally, maize, beans and groundunts. The rest of those catchments were cleared for cattle grazing. The farmers had to adopt soil conservation methods. Two other catchments were left as forest. The hydrologic conclusions were that a significant increase in runoff occurred on the agricultural catchments and the shape of the hydrographs changed. The peak discharges nearly doubled, and the time to peak and time to base decreased, which indicates that the agricultural treatment generates quicker flow.

In Nepal (monsoonal), it has been claimed that local deforestation decreases infiltration and increases flooding in the highly populated areas downstream. Gilmour et al. (1987) measured hydraulic properties in a range of vegetation treatments. They found that reforestation did increase hydraulic conductivity, but values were so high under any treatment that all but the heaviest of monsoon events were within the capacity of the soil to absorb. This is a good example where precise measurement redefines an issue!

Early, explicit data from 1938–1967 on the influence which afforestation of an agricultural watershed has on runoff (in the Appalachian mountains in Ohio, USA) are given by Ricca et al. (1970). There was a reduction of runoff over the first 10 years and a levelling off for the following 15 years. They infer from declining groundwater levels that percolation potential decreased with time, and that, over the last 10 years of the work, hydrologic stability was achieved.

Sartz (1973) reports the effect of forest cover on soil freezing and overland flow. He observed that the forests in Wisconsin rarely yield overland flow, even when the ground is frozen. It is not clear why water infiltrates frozen forest soil yet runs off frozen non-forest soil. This paper demonstrates the importance of litter and the canopy on freezing depth and overland flow. The four treatments were: wood vegetation cut and removed; all vegetation cut and removed; uncut forest with litter removed; undisturbed forest (Table 7).

How would you have predicted the outcomes of these treatments? Removal of litter and removal of all vegetation increased freezing depth and overland flow. Removal of woody vegetation only decreased both the foregoing parameters. The increases in freezing depth and overland flow appeared to be related to changes in soil bulk density and porosity. Sartz asked of his readers not to infer from the results of this study that normal logging operations, or even prescribed burning, would also produce comparable amounts of runoff. Small research plots may not, he warns us, indicate accurately what would happen on larger areas!

Table 7. Mean frost depth after removal of litter, of all vegetation and of woody vegetation, compared with undisturbed forest[a] (After Sartz 1973)

Treatment	1970	1971	1972
		—cm—	
Litter removed	3.6	9.1	39.1
All vegetation removed	1.9	4.7	22.6
Woody vegetation removed	1.4	0.1	8.8
Undisturbed forest	1.9	1.4	12.4

[a] All values of 1971 and 1972 are significantly different from each other as determined by T-tests.

Two papers evaluate traditional drainage in afforestation in "Europe". Savill and Weatherup (1974) found no effect of afforestation on water runoff in Ireland. They were concerned with yield of water to reservoirs. Prior to planting, the forest area had been artificially drained. Trees were planted on the drain ridges which were expected to induce faster runoff. Further, the forest soil proved to be less permeable than the grassed control areas. The authors frankly present the weaknesses of the experiment – different plot size, lack of prior flow measurement, catchments not isolated – thus allowing the reader to interpret and evaluate.

The second paper, by Robinson (1986), concerns the necessity of surface drains in northern Europe in establishing plantation forestry. The drains increase annual runoff, shorten flow response times and increase the peaks. Robinson notes that, as the trees mature, the magnitudes of the foregoing effects decrease due (I suppose) to the greater interception and to roots colonizing the drains and impairing their hydraulic efficiency.

Verry et al. (1983) review the effects of clear-cutting in central Minnesota in regions of snow, which tends to complicate the storm-flow characteristics. They cleared part of an aspen forest and followed peak discharge, storm-flow volume and the timing of peak discharge. They evaluated snowmelt and summer rainfall runoff events.

Clearcutting caused snowmelt peak discharge to increase by 11–143% and rainfall peak discharge to increase by up to 250% during the first 2 years, before declining toward precutting values. Storm-flow volumes from rain increased up to 170% but declined to precutting volumes in the third year. Snowmelt volumes did not change size but peak discharge occurred 4–5 days earlier after clearcutting. There was no change in timing of rainfall-induced storm flow.

Wildfire, an agent of deforestation, may induce extensive runoff and sedimentation. Helvey (1980) documented the increase in runoff for one forest fire in north-west USA.

In the Mediterranean climate of south-western Australia ($1000 \, mm \, yr^{-1}$ rainfall), Williamson et al. (1987) present numerically and graphically the effects of clearing on direct runoff and on storm hydrographs in paired catchments (Table 8). The latter show large differences after clearing. Storm runoff from the cleared catchment is seven times, and over the whole year it is three times, the control value.

Table 8. Effect of forest clearing on direct runoff (QR) and on total stream flow (QT) as revealed by a paired-catchment study in south-western Australia. The cleared catchment was subjected to pre-clearing calibration (After Williamson et al. 1987)

Hydrograph analysis type	Catchment	Total stream flow Q_T/Q_{TC}		Direct runoff Q_R/Q_{RC}		Direct runoff/ total Q_R/Q_T	
		Pre	Post	Pre	Post	Pre	Post
Storm	Cleared	0.88	6.44	0.89	18.8	0.59	0.82
	Control					0.58	0.32
Full year	Cleared	0.87	2.45	0.84	12.4	0.04	0.16
	Control					0.04	0.03

Sharma et al. (1987b) examined the spatial distribution, dependence and variability of sorptivity (S) and saturated hydraulic conductivity (K_s) on forested and cleared catchments in south-western Australia. Although the lateritic soils have very high infiltration rates, within 2 years after deforestation and conversion to agriculture, K_s was retarded ten times and S by a factor of three. Despite this fall, these rates are still sufficiently high that there is little probability of direct surface runoff. The dominant mechanism of runoff generation on such farm land appears to be saturation of surface soils through subsurface flow due to enhanced recharge after clearing.

Liu (1987) followed the change in runoff (and also flooding) in north China as the forests were used for industrial energy. He analyzed twenty three drainage basins ranging in area from 150–1500 km² and with proportions in forest cover ranging from 2–98%. He had 30 years of records available. With equal rainfall amounts, peak runoff from the wooded basins is much lower than from unwooded basins. Flood peak discharge and flood frequency were analyzed for one of the basins of 1200 km² which had been deforested. Values for peak discharge ($m^3 s^{-1}$) were 680 under forest, 870 with mixed vegetation and 960 unforested. Flood frequency curves show, particularly for larger floods, that the frequency was a little greater.

Finally, an early paper by Nakono (1967) describes four natural and artificial forests in Japan that have been clear-cut or destroyed by typhoons. He records the change in runoff up to 22 years after "clearing", and concludes that runoff was greatest in the year after clearing and was eventually fully restored; that increase in runoff depends on topography; that runoff from snow increases after clearing; and, in general, runoff was 28–58% greater and peak flows increased by 50–114% immediately after clearing.

4.3.2 Deforestation, Revegetation and Water Yield

This section is at the heart of this review. It brings together some 50 papers on the effects of deforestation and revegetation of catchments on their water yield. I start with some remarkable integrating papers which are unique in hydrology. Specific papers follow to assist those readers who seek answers to local problems in specific climates. My brief reporting of them serves as an index of interests over a wide range. Some unexpected, perplexing issues arise.

The first modern reviewer of forest treatment on water yield was Hibbert (1967). He reported 39 studies which, collectively, revealed that deforestation increases water yield and that reforestation reverses the trend. Today, we take that for granted. He also found that the response to treatment was highly variable and, for the most part, unpredictable.

Bosch and Hewlett (1982) increased the sample of 39 above to 94 catchments. They assessed the extent of percentage change in vegetation cover on water yield for different classes of vegetation. Only one catchment failed to show an increase in yield. The approximate magnitudes of changes from their sample were: pine and eucalypt types caused an average change of 40 mm yr⁻¹ in water yield per 10% change in cover, whereas for deciduous hardwood and scrub the respective value was only 25–10 mm yr⁻¹.

Bosch and Hewlett were aware of the statistical bluntness of their findings. They suggested how inferences might be made more confidently if further selections of samples were to include a greater range of rainfall regions.

Trimble and Weirich (1987) and Trimble et al. (1987) concentrated on reforestation and its ability to reduce flow. They studied ten large river basins (in south-eastern USA) because treatment effects are easier to discern on a large scale.

The regressions of annual stream-flow change compared with percent of basin reforested or deforested are plotted together with the catchments reviewed earlier by Bosch and Hewlett (1982). The ten large river basins, when added to the 55 of Bosch and Hewlett increases R^2 from 0.38 to 0.50. Trimble and Weirich concluded that:

1. Relatively small increases in forested land (10–28% of total basins) within populated stream basins significantly reduced water yields over large areas and long time periods, thus empirically corroborating and extending the results of 70 years in small basins;
2. The additional loss of water due to forest cover is about 0.3 m^3 per m^2 of land reforested.

Now follow specific examples of the relation between vegetation and water yield. Each is a précis of the circumstance and outcome. You will see again how unsatisfactory it is to generalize.

Chang and Watters (1984) were concerned with water yield from forests in east Texas which has a humid subtropical climate and a precipitation of about 1100 mm yr^{-1}. The forest region occupies 93×10^3 km^2. The proportion of land as forest in the 19 unregulated catchments of the study area averaged 54% with a range of 5–97%. They analyzed the effects of proportion of watershed in forest and several physioclimatic factors on mean annual stream flow (1965–76), median flow and 12 flood-flow characteristics.

Annual stream flow increased with decreasing area under forest with a difference of 200 mm yr^{-1} between extremes. Of the 31 physioclimatic parameters analyzed, the most significant variables in affecting stream flow characteristics were watershed area, percent under forest, shape index, spring precipitation and annual temperature. Using two to three of those five variables, all of the 14 stream flow characteristics can be estimated with accuracy ranging from acceptable to excellent levels. How many readers could make use of such an approach or are already doing so?

Perhaps without fully realizing it, the catchment manager may play an important part in developing water resources through his harvesting of forest products. A less obvious influence is in the choice of tree species to be grown. So wrote Swank and Miner (1968) after they had reduced water yields by converting a forest from hardwoods to white pine (*Pinus strobus* L.). Their site was in western North Carolina, precipitation 1900 mm yr^{-1} (2% snow) and humid, subtropical climate. Once the crowns of the pines began to close, stream flow steadily declined at a rate of 25–50 mm yr^{-1}, and after 10 years it was 100 mm yr^{-1} less than from the original hardwood forest. They attributed this loss to greater interception from pines.

Baker (1986) explored tree clearing in the ponderosa pine region of Arizona with a view to increasing water yield (Table 9). Precipitation ranges from 580–740

Table 9. Significant increases in annual predicted water yield over varying reductions in basal area of ponderosa pine forest in Arizona, USA. Values are given over several successive years after treatment (After Baker Jr. 1986)

Years after treatment	Overstorey removal			Strip-cut with thinning		
	100%	77%	33%	68%	57%	31%
	(mm)	(mm)	(mm)	(mm)	(mm)	(mm)
1	62	63	70	72	33	25
2	56	58	61	62	27	25
3	50	53	52	51	22	25
4	44	48	42	n.s.	17	25
5	38	43	32	n.s	n.s	
6	32	38	23	n.s	n.s	
7	27	33	n.s	n.s	n.s	
8	n.s	28	n.s	n.s	n.s	
9	n.s	23		n.s	n.s	
10	n.s	18		n.s	n.s	

mm yr^{-1} and occurs in summer (200–300 mm) and in winter (350–500 mm). There is a significant snow component which is influenced by clearing treatment.

Annual water yields were determined for three levels of overstorey removal and three levels of strip cut with thinning applied on ponderosa pine catchments. Water yield increases from a completely cleared catchment were statistically significant for 7 years, after which the growth of Gambel oak and herbaceous vegetation had compensated. The duration of effects from the other treatments is presented, including the influence of slope aspect. Baker concludes that the potential for increasing water yield in ponderosa pine is less than from other commercial forest types because it occurs on drier sites.

McGuinness and Harrold (1971) in Ohio (continental, warm summer climate) sought answers to three questions concerning reforestation and stream flow:

1. Are flow duration curves useful in assessing hydrologic effects of reforestation? Their data indicated that they are useful.
2. As periods of flow duration are made shorter, at what point does the effect of reforestation in reducing maximum annual flow become insignificant? From the 12th year onwards;
3. Has reforestation delayed the date in autumn when stream flow normally increases in volume? Their answer is – yes.

They took a catchment of 18 ha (950 mm yr^{-1}) on 70% of which pines (unspecified) had been planted in 1938. The remaining 30% was left in hardwoods of differing ages. By 1945, complete canopy cover had occurred. This was paired with an agricultural catchment of 120 ha of which 30% was under woodland. Measurements continued untill 1968. In their general conclusions they point to the necessity of operating experiments of this kind over a sufficient period for gradual trends to become apparent. Had they stopped at 1957, they claim they would have obtained less insight into hydrologic effects of vegetation.

In western Oregon (marine, west coast climate, precipitation $2300\,mm\,yr^{-1}$), Harr and McCorison (1979) reported the initial effects of clear-cut logging of conifers on the size and timing of peak flows. Their findings were presented with respect to reduction by snow in Sect. 4.1.2. In contrast, they found no significant changes in size or timing of peak flows that resulted from rainfall alone. They were unable to account for that. The sharp difference between snow and rain is shown in a pair of hydrographs typical of the post-logging period. In one, the first part of the storm was rain and the response was rapid. Then it snowed and the rate of the hydrograph rise decreased almost immediately. In the other hydrograph, snow preceded rain and the reverse is immediately apparent.

In a nearby location, Rothacher (1970) had found that increases in water yield following timber harvest roughly conform to the proportion of the area cleared. In the high precipitation areas of the Oregon Cascades mountains, clear-cut logging can increase annual water yield by more than the 450 mm suggested by Hibbert (1967). Approximately 80% of that increase occurs during October to March.

In the subalpine conifer forests of Colorado (elevation 2900–3500 m), Troendle and King (1987) examined the hydrologic response of two treatments. These are the so-called patch clear-cut and the partial cut with each treatment removing about 40% of the stand. Patch clear-cut was defined as commercially clearcutting 12 small units uniformly spaced through the catchment. These were circular openings 122 m (or 5 × tree height) in diameter. All trees were felled and the undergrowth cut to a height of 10 cm. The rationale behind such geometry was to maximize snow pack accumulation in the clearcuts and to optimize flow increases for the basal area removed.

Annual and peak flows from the clearcut catchment were increased significantly. The partial cut gave a significant increase in total water equivalent in the winter snowpack and an apparent increase in total annual stream flow that was comparable to the clearcut. The precision of their data is splendid!

Ffolliott et al. (1989) present a position paper on the ability to increase water yield from the snowpack by deforestation techniques in Arizona and New Mexico. A simulation model was developed called SNOW. It deals quantitatively with practical management issues starting with reducing forest densities and finishing with downstream reservoir regulation.

Hibbert and his collaborators published three papers, over a 14-year interval, on the advantages of changing the nature of vegetation for increasing water yield. In the first (Hibbert 1969), a hardwood forest in western North Carolina ($1800\,mm\,yr^{-1}$) was cleared and planted with Kentucky 31 fescue grass which was neither grazed nor harvested. In the season of high grass productivity, the catchment yield of water was similar to that for the prior forest. In the season of declining growth the reverse held. His main finding was the strong inverse relationship between stream flow and grass. It follows, I suggest, that had the grass been commercially grazed or harvested, there would have been a substantial increase in water yield.

Hibbert et al. (1975) investigated the effect of large-scale conversion of the chaparral vegetation association in Arizona ($400\,mm\,yr^{-1}$) in a further investigation of increased water yield. From this widely based and thorough investigation they estimated the potential of large-scale conversion to grass. While not describing the detailed results in this paper, I shall set out his economic feasibility studies and general constraints derived from Brown et al. (1974), because they have a strong bearing on whether a hydrologically successful revegetation program should be undertaken or abandoned. The study method was as follows:

1. Delineate chaparral areas meeting certain criteria for conversion.
2. Estimate costs of conversion.
3. Determine effects of conversion of yields of water, forage, and sediment, as well as impacts on fire hazard, recreation use, aesthetics, and wildlife habitat.
4. Develop value criteria for quantifiable effects and impacts.
5. Integrate measurable costs and benefits into an economic framework using benefit-cost analysis.

Of the 340 000 ha available, 200 000 ha were initially rejected because they were either in wilderness areas, had less than 30% shrub cover, or the slopes were greater than 60%.

Hibbert (1983) then extended his ideas on management of water yield by changing the vegetation of the western rangelands of USA based on the "world wide" review of the 94 catchments by Bosch and Hewlett (1982, this section). He sets out five criteria to be met which are, in effect, a set of hydrologic conclusions for this section:

1. Precipitation should exceed 450 mm yr^{-1} (Bosch and Hewlett 1982).
2. Maximum efficiency occurs where precipitation is concentrated in the cool season. It may be possible in a cool climate to get response at less than 450 mm yr^{-1}.
3. Replacement vegetation must use less water.
4. Replacement species should be of low biomass, deciduous and shallow rooted.
5. High water-use plants should be thinned or eradicated.

To return to the main thrust of this section, Hornbeck et al. (1970) and Hornbeck (1975) at the Hubbard Brook Experimental Forest, New Hampshire, USA, conducted an extreme experiment on forest clearing and water yield. Their objective was to obtain maximum stream flow during the summer low-flow period by eliminating transpiration. Following clearing, they prevented regrowth with herbicides. They were rewarded over the first 2 years with an increase of 300 mm yr^{-1}, most of which occurred during the low-flow months of June–September. There was also a small increase in snowmelt runoff and a consistent increase in the high flow rate of the "growing season". They concluded that substantial increases in stream flow can be induced in the uplands of eastern USA. The authors were fully aware of the impracticable and harsh treatment. It enabled them to assess economic considerations.

Ponce and Meiman in their (1983) paper discussed: (1) the prediction of increased yields from large basins; (2) economic evaluation of additional flows; (3) court acceptance and need for system models; (4) the legal question of ownership and transferability of increased yields, and (5) management emphasis on private and federal lands.

The immediate potential for water yield augmentation is on carefully selected watersheds that have the biophysical potential to produce high value water under environmentally acceptable multiple-use management. They predict that water yield management on a broader scale will result from increased pressures to solve the legal and economic issues involved.

The foregoing paper induced a Discussion Paper by Turner (1984). He took up the issues of evaluation of increased flows, system models and legal problems. Of models he considers that the errors in stream flow and precipitation data will be larger than the amount being salvaged at times. Under legal problems he discusses the rights of storage and other matters in such a succinct style that all readers who wonder about these matters will appreciate his exposition.

Not all vegetation changes which alter stream flow are deliberate. In the paper by Black (1968), the change in vegetation was caused by farm abandonment in New York State (continental, warm summer climate) between 1921 and 1961. Vegetation on farms passed through a brush and scrub forest stage before reverting to the mature forest. Which direction, do you suppose, did the change in stream flow take? It increased!

It was established that precipitation and estimated total evaporation had not changed. What caused the increase in the proportion of precipitation appearing as runoff, particularly in the dormant season? One possibility was increased urbanization in the catchment, but this could be discounted. A second hypothesis concerns the cause of a delay of 10 days in runoff. The author claims that the slight lengthening of the shortest one-quarter flow interval, the time of year during which the increase occurs and the nature of the vegetation suggest that increased snow accumulation, shading and prolonged melt are the causes of increased runoff. You may disagree. Yet, it is clear that effects of vegetation on the hydrologic cycle are sometimes subtle beyond the expectation of scientists and engineers – a point to be made occasionally.

Further to Black's paper, Douglass (1983) makes a shrewd point about the potential for water yield augmentation by forest management. An important restriction is not the availability of models but the land ownership patterns and the economic objective of owners.

A further issue is the development of methods for predicting increases in water yield in relation to timber harvesting and site conditions. Galbraith (1975) predicted that within the next 10 to 20 years (now) the remaining areas of wild forest land not reserved for wilderness or other undeveloped uses will come under management in most areas of the National Forest System. As the era of intensive forest management draws closer, he argues, so also does the need to evaluate more precisely the response of watersheds to increasing use. Fundamental to any assessment of watershed behaviour, whether it be surface erosion, slope stability, or channel dynamics, is the ability to predict increases in flow due to removal or modification of the vegetative cover.

My review of the influences of vegetation on stream flow and water yield in North America ends dramatically:

"A rushing, wild wind swept across the high plateaus of Colorado in 1939. In its wake, a multitude of bark beetles thrived and multiplied – first in the weakened, wind-thrown, mature Engelmann spruce, then in the standing green timber. Within seven years, the beetles had destroyed 4 billion board feet of standing timber..., and left large areas... resembling a lifeless, ghost forest. Successive, massive flights flew northward and eastward... where they repeated the destructive exploits of their forebears."

The unlikely publisher of this drama is the Journal of Hydrology! Bethlahmy (1974) reported this beetle epidemic which destroyed the timber in two large drainages but bypassed a third drainage. Long-term stream-flow records were available for those drainages for the periods before and after the onset of the epidemic, analysis of which showed that it was responsible for a major increase in yield.

The latter part of this North American literature review reveals that deforestation takes many forms other than those arising from resource management.

The south-west region of Australia has a Mediterranean climate. Prior to European settlement the main vegetation was eucalypt forest. In their primal state the ground flora and middle storey were sparse, but after harvesting of timber, a new ecosystem developed in which the middle storey and ground flora became denser. This had profound implications for land management as I shall make clear.

In the higher rainfall areas near the west coast ($> 900 \, \text{mm yr}^{-1}$), the forest is used for timber and water supply. In the lower rainfall areas ($< 900 \, \text{mm yr}^{-1}$, further inland) it has been largely cleared for agriculture. Forest management is faced with a multiple dilemma:

* For water supply, total evaporation from the forest needs to be minimized, which implies thinning;
* Thinning trees for water harvesting stimulates undergrowth which in turn increases evaporation (Greenwood et al. 1985b). While it is practicable to thin trees, it is impracticable to thin undergrowth and leave the trees;
* For seasonal crops and pasture, farmers prefer complete clearing of the forest. This system greatly reduces evaporation (Greenwood et al. 1985a), which increases recharge of groundwater which in turn rises into a horizon of primeaval cyclic salt. The salt is then mobilized and destroys the crops and pollutes the watercourses which eventually drain into storage catchments located in the high rainfall area.

The following three papers work through partial resolutions of the above dilemma. The paired-catchment paper by Williamson et al. (1987) has already been presented in Sects. 4.2 and 4.3.1. Water yield increased immediately after clearing. In the higher rainfall zone, stream flow increased about four times to 48% of rainfall. Absolute increases in stream water from partially cleared catchments in the lower rainfall location were about an order of magnitude less than for the higher rainfall catchment, though the relative increases were similar.

Ruprecht and Schofield (1989) examined the mechanisms of stream-flow generation following total or partial deforestation of catchments. Their evidence is drawn from the catchments of Williamson et al. (1987) above, and from numerous other experimental catchments in the region. Agricultural development eventually increased stream flow by 30% rainfall per year. In the first year the increase was only 10% of rainfall, but it continued to increase at a slower rate for a further 5 years when a new equilibrium was reached.

Explanations of the above time trend of stream-flow increase were sought in the mechanisms of stream-flow generation. The initial increase was attributed to the impact of the immediate loss of interception (13% of rainfall). The subsequent linear increase in stream flow was closely correlated with the expansion of the groundwater discharge area. The cessation of stream-flow increase was considered to be due to the attainment of a new groundwater recharge–discharge equilibrium. Evidence from other catchments, which have undergone forest reduction, shows that the permanent groundwater system is instrumental in controlling stream-flow response following deforestation.

The work of Stoneman and Schofield (1989) was prompted by the increasing cost of supplying metropolitan water because all the readily available forested catchments had been utilized (Fig. 7). They were also mindful of the predicted

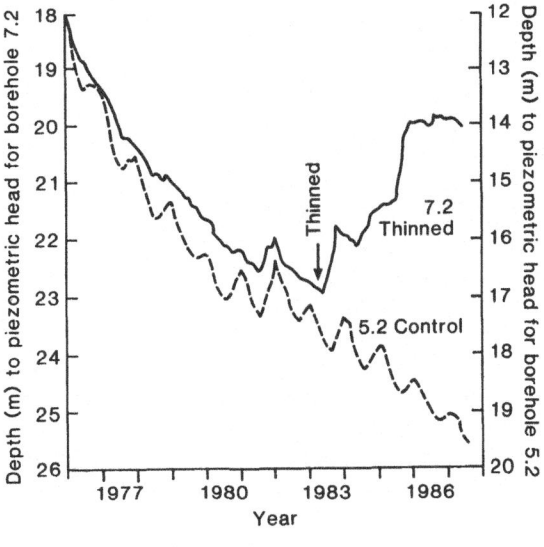

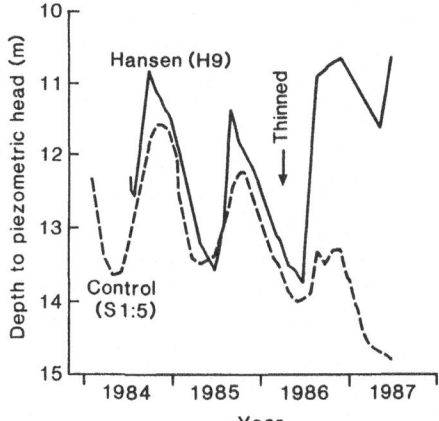

Fig. 7. Effect of thinning of the jarrah forest, south-western Australia, for water production in two catchments. (After Stoneman and Schofield 1989)

reduction in stream flow from the "greenhouse effect". In exploring the value of forest thinning, they produced four indirect estimates of the increased yields of water from thinning the forest. The value of the estimate is not as important here as are the four approaches they devised, which were estimates from: experimental catchments such as those of Williamson et al. (1987); canopy-cover stream-flow relationship; water balance changes; model simulations.

In surveying the above analyses, the authors consider that, although each individual estimate has a relatively low reliability, all methods estimate substantial increases in stream flow. The most conservative estimate for both the high and intermediate rainfall zones was 11.5% of rainfall.

To the south of the jarrah forest lies the extensive karri forest association (*Eucalyptus diversicolor*), rainfall $> 1100 \, \mathrm{mm \, yr^{-1}}$, and also in the Mediterranean climatic zone. What is the quantitative change in water yield following logging operations and subsequent regrowth? Borg et al. (1987) measured this on four catchments, and Stoneman et al. (1988) quantified the vegetation cover before and after logging. (Their methods of vegetation survey should be read by all forest hydrologists).

The range of forest conditions was represented by four small catchments. Logging was a commercial exercise. Regeneration began 1 to 30 months after logging. Stream flow increased and groundwater levels rose for 2 years and then gradually declined, though somewhat obscured by variation in rainfall. After about 12 years of regeneration it is expected that stream flow will return to the pre-logging values. The overstorey cover reached the unlogged density after 10 years. It stabilized at a higher value than is typical for unlogged stands. Total vegetation cover reached the unlogged value within 5 years, rose for 5 more years and also remained above the unlogged value.

The rigorous surveying of natural, as distinct from plantation, vegetation cover has not often accompanied catchment experiments on deforestation. What is the value of such surveys? Suppose we can relate water yield to vegetation cover in representative catchments. How shall we apply it in practice? If we aim to increase yield by a certian amount, we estimate the logging intensity which will induce that increase. Yet, we cannot expect to hold that constant because recovery begins immediately. It would not be feasible to attempt regular small-scale logging to hold vegetation constant. That is not to belittle those who carry out such surveys as a scientific exercise in explaining the controlling cause of an effect, as in the foregoing example.

In a temperate-zone catchment of eastern Australia (rainfall $900 \, \mathrm{mm \, yr^{-1}}$), Aston and Dunin (1980) modelled the effect of afforestation of grasslands on water yield. Their model is excellent in its simulation of measured water balance parameters. Afforestation under pines was predicted to reduce stream flow by 32%. The model is described explicitly.

Gilmour (1977) examined the effect of logging and clearing on water yield in a $4040 \, \mathrm{mm \, yr^{-1}}$ site in north-east Queensland, Australia. A pair of small experimental catchments was used to determine the effects of logging and clearing on water yield in a typical lowland rainforest area. Annual water yield increased by 10% during the first 2 years after clearing, whereas the prior logging operation was almost without any hydrologic effect. This, Gilmour concluded, is because the rainforest yields a very low volume of timber compared with most other commercial forest types. Even after heavy logging, a large volume of timber and a large canopy remains, implying high total evaporation rates. Flood flows appeared to be determined more by rainfall characteristics and catchment geomorphology and its wet or dry condition than by the vegetation changes.

The drama of the hydrologic beetles in USA in 1939 (this Sect.) is upstaged by the bushfires of south-eastern Australia, also in 1939, and again captured by the Journal of Hydrology (Langford 1976). Since forest fires are of widespread occurrence, it is important to obtain some quantitative evidence of their effect. The fire caused complete defoliation instantly so that one would expect greater yield of water after the commencement of the wet winter season (rainfall range 1200–1800 mm yr^{-1}). Unfortunately, Langford rejected the stream flow in the 3 years immediately after the fire in the interests of long-term evaluation of trends. After those 3 years, water yield had declined significantly and over the next 15 years, even more rapidly. The data came from four burnt catchments matched against one that escaped the fire. The reduction in flow reached 24% of average pre-fire flow. The cause of that reduction is pure forestry.

The dominant tree species is the tall (100 m) *Eucalyptus regnans*. It has a clean trunk and minimal canopy, implying low interception and transpiration. After the fire, evaporation would be zero. But the germination of seeds is so dense and its juvenile growth so fast that its leaf area per unit ground area soon exceeds that of the prior canopy. This accounts for the reduction in water yield. The status of forest hydrology would have been enhanced if water engineers had explored the mechanism of plant hydrology so effectively as did Langford.

Brown (1972), also in south-eastern Australia, reported the immediate re-establishment of gauging stations after a fire and so was able to compare, on two catchments, the hydrologic effects of fire over several years. The catchments had been covered with wet sclerophyll and dry sclerophyll forest at the lower levels, and open alpine woodlands at the higher elevations. The climate is temperate with significant rain in each month, and with some snow in winter. The main conclusions, supported by rigorous data, were:

1. Occurrence of sharp secondary peaks on the rising side of many flood hydrographs of one catchment;
2. The magnitude of flood discharges from the other catchment was greater than before the fire with the highest flood in the 19 years of record occurring 11 months after the fire as a result of rainfall of relatively low intensity;
3. Runoff on both catchments was significantly greater than it would otherwise have been for at least the first 4 years after the fire;
4. Measured sediment loads were very much greater;
5. There was a progressive reduction in all the above effects with the passage of time.

Regrowth within the catchments 6 years after the fire was almost 100%. Several years of low rainfall may have slowed the process.

In a multiple catchment experiment in South Africa, Van Wyk (1987) assessed the influence of afforestation of grassland with *Pinus radiata* of stream flow for the 40-year cycle from planting to the felling of one catchment. It was intended to resolve the controversy as to whether stream flow is adversely affected by extensive timber plantations. The effects are evaluated against an unafforested control. The climate is Mediterranean with 85% of the rain falling between April and September. There is a rainfall gradient across the study area ranging from 2300–1300 mm yr^{-1}. The percentage afforestation of each catchment ranged from 98 to 36%.

Afforestation reduced stream flow. In the case where 98% of the catchment was afforested, stream flow decreased by 313 mm from the initial 663 mm yr^{-1} over a period between 12 and 32 years after afforestation. Stream flow stabilized at that level. For the catchment with 57% afforestation, stream flow declined by 200 mm yr^{-1} from 593 mm yr^{-1} over the period from 16 to 40 years after afforestation. Here, stream flow stabilized after about 20 years. The actual detailed time trends are explicitly illustrated. The conclusion is that afforestation with pines strongly reduced stream flow, but as only 7% of the region is afforested, the total water supply is not greatly affected. Presumably, it is not intended to increase the area of afforestation.

Van Lill et al. (1980) reported the long-term effects of afforestation on flow in a region of summer rainfall in the Transvaal of South Africa (rainfall about 1200 mm yr^{-1}). After a calibration period of 12 years, *Eucalyptus grandis* was planted in one catchment, a second was planted to *Pinus patula* 2 years later, and a third was maintained under natural grass. Regression analyses showed that *E. grandis* reduced flow from the third year after planting with a maximum reduction of between 300 and 380 mm yr^{-1}. Maximum reductions in seasonal flow were 200–260 mm yr^{-1} in summer and 100–130 mm yr^{-1} in winter. Conclusions for *P. patula* were as yet tentative and I have not found a later publication confirming whether *P. patula* was less effective than *E. grandis*.

In the central North Island of New Zealand, Dons (1986) reported the long-term decline in the flow of the Tarawera river due to afforestation of poor pasture, gorse and indigenous scrub land with pines. Between 1964 and 1981 the annual, summer and winter flows were reduced by 10.9, 11.4, and 9.6 m^3 s^{-1}, respectively. Simple flow models of two neighbouring unforested catchments showed that about 4.5 m^3 s^{-1} of these reductions, or 13% of the mean flow over the calibration period, could be attributed to afforestation, while the remainder was due to decreased rainfall.

Two papers by Calder display the discipline of hydrology at its best in dealing with controversy. One (1986a) concerns the afforestation of upland Britain which threatens water yield of important catchments. The other (1986b) is the fierce outcry from southern India that eucalypt plantations are exhausting the groundwater supply. In both, he takes a speculative and inferential view for he is addressing problems which he knows how to resolve but lacks the data to do so. He proposes the rigorous data to be acquired as a first step in dealing with those controversies.

Hsia and Koh (1983) measured the water yield increase resulting from clearcutting a small (6 ha), low altitude experimental basin in Taiwan. The basin was covered originally with subtropical montane hardwood forest and ground disturbance was kept to a minimum as skyline logging was used. The climate of the studied area is characterized by a distinct wet (from May to September) and dry season. Rainfall during the wet season averaged 1680 mm or 82% of the annual rainfall. Water yield increases of 402 (55%) and 184 mm (47%) were estimated for the first and second wet seasons after the clearcut, respectively. For two dry seasons the yield increases were 46 (108%) and 20 mm (293%). Although the ratios of increase were high during the dry seasons, the absolute amounts accounted for only 10% of the total annual yield. The results suggested that water yield increases after small-scale deforestation had little effect on the water supply during the dry season.

One might assume from this section of the review that deforestation leads almost universally to increased water yield, that is until one confronts the Russian literature, reviewed by Rackhmanov (1970).

Rackhmanov acknowledges that the international research on small catchments has reported high evaporation rates from forests, decreased stream flow with growth of forest cover, and the reverse for cutting and burning of forests. Investigations on small watersheds in USSR produce the same conclusions. For *large* river basins in USSR, and based on data from standard network-design observations, annual stream flow clearly *increases* under the influence of forests. This phenomenon I call "the Russian paradox".

First, Rackhmanov discounts pre-1950 Russian work because of its inadequate data base. Then he uses the network of hydrometeorologic stations on large flat areas of afforestation in order to minimize topographic variance in precipitation and runoff. Rakhmanov's data seem as convincing as any other data we have been reviewing. Coefficients of correlation are high (> 0.80, $P = 0.005$) of paired basins ranging from 500 to 12 000 km^2.

Rackhmanov's explanations of the positive effects of forests on stream flow are:

1. Increased precipitation occurs over large forests (measured);
2. Reduced evaporation from forest terrain (conjectured) due, he says, to reduced air temperature, low wind velocity, increased air humidity, protecting effect of forest litter and greater accumulation of snow.

These explanations require confirmation. Fortunately, Garczynski (1980) took up the issue. Following clearfelling to maximum areas of 700 ha in a single block, the annual discharge increased from 40 to 400 mm, from which USA workers have concluded that forest cover produces less runoff than grass cover. Russian workers, he affirms, on the basis of correlations between annual stream flow and percentage forest cover of basins 10 000 km^2 and more, found the opposite true for basins over 10 km^2. USA hydrologists have found the same correlation in 140 basins of less than 260 km^2 in the northeast yet continue, he claims, to be sceptical.

Multiple correlations on these data and also from Oregon and California, USA, show that when channel length is more than 15–30 km (depending on climate), the influence of forest on runoff becomes positive, or at least ceases to be negative, and affects mainly the base flow. This increase cannot be explained by the slight increase in precipitation and must, considering evaporation, have other causes. If these results can be proved correct elsewhere, they should modify the most recent concept concerning the influence of vegetation on the hydrologic regime.

Garczynski then develops a "first" explanatory hypothesis. Forest favours maximum infiltration of precipitation which is thus temporarily lost to the stream but reappears further downstream in the channel. He then speculates on recycling of precipitation in forests and on the levels of evaporation within and outside forests.

The next sections on soil water storage and on recharge will also impinge on the Russian paradox.

4.4 Storage of Soil Water

There have been many crop and forest water balance projects which have measured the changes in the storage of soil water under crops and forests with time. Few have been devoted to comparing storage under forests and crops simultaneously. Some examples follow.

Sharma et al. (1987a) present a 5 year study in south-western Australia (Mediterranean climate) which was based on two sets of "matching" catchments: one set in a rainfall of $1200 \, mm \, yr^{-1}$ and the other of $800 \, mm \, yr^{-1}$. After 3 years of testing for uniformity, one of the 1200 mm catchments was clearfelled and one of the 800 mm catchments had 53% of its area cleared. The cleared land was sown to annual pasture. Soil water storage was measured down to 6 m with a neutron probe in 15 access tubes located at five stratified sites in each catchment.

Large spatial variability in soil water storage was encountered within a site, between sites within a catchment, and between paired catchments; the dominant variability being between sites. Comparisons between the pre- and post-clearing states within a catchment and between the cleared and uncleared control catchments were used to evaluate the effect of change in land use on soil water storage.

Within 2 years of the change from forest to pasture, a significant increase in water storage had occurred in the profiles in both cleared catchments. Concurrently, there was a small decrease in the uncleared control catchments. The increases following clearing were greater in the higher rainfall catchment than in the lower rainfall catchment, more pronounced in the first year than in the second year, and occurred mostly at depths greater than 2 m. In the 1200 mm catchment, the increase in summer minimum storage in the first and second years amounted to 220 and 58 mm, respectively, whilst for the 800 mm catchment the increase for the first year was < 50 mm. The increased storage was due to a lower leaf area duration from the shallow-rooted, wet season pasture which extracts water from the top 1 m or so. The evergreen eucalyptus forest extracts water from depths down to 6 m and beyond.

In humid east Texas, USA, Chang et al. (1983) comprehensively measured soil water regimes under a range of silvicultural treatments based on the following rationale (Table 10): in the southern forests of the USA, heavy equipment removes all vegetation and slash to facilitate replanting and to conserve soil water for the establishment of the new rotation. The risk in this procedure is that the bare soil is highly erodible, due mainly to surface runoff. When soil water is below field capacity and the rainfall intensity is less than the infiltration rate, much of the rain enters the soil and runoff and erosion are minimized. Thus, soil water content is of key importance in storm runoff and erosion. What then is the procedure to adopt to minimize storm runoff and erosion?

The forest site chosen had a mean annual precipitation of 1950 mm. The main climatic features were frontal storms in winter and convective afternoon storms of short duration, high intensity and low frequency, in summer. The forest species were 40-year-old loblolly (*Pinus taeda*) and short-leaf pine (*P. echinata*) with scattered

Table 10. Effects of vegetation and surface treatments on mean antecedent soil moisture, net rainfall, surface runoff and sediment of 33 runoff-producing storms over 2 years in Texas, USA (After Chang et al. 1983)

Treatment	Net rainfall (mm)	Runoff (mm)	Sediment (kg ha^{-1})	Antecedent soil moisture (g cm^{-3})
Forest	21.8	0.64	0.426	0.238
	(2.62–159.54)	(0.0–12.64)	(0–4.30)	(0.110–0.368)
Thinned	22.9	1.44	0.889	0.306
	(3.22–164.14)	(0.0–14.04)	(0–8.50)	(0.160–0.405)
Commercial	25.6	2.05	6.20	0.297
cutting	(4.04–179.16)	(0.0–14.49)	(0–111.30)	(0.096–0.474)
Chopped	26.8	4.81	18.55	0.345
	(4.4–186–44)	(0.09–38.16)	(0.1–133.40)	(0.174–0.461)
KG-bladed	26.8	7.12	143.26	0.341
	(4.4–186.44)	(0.1–54.71)	(0.1–979.10)	(0.247–0.527)
Cultivated[a]	29.5	14.09	349.49	0.423
	(4.4–186.44)	(0.1–152.85)	(0–1395.60)	(0.329–0.518)

[a] Based on data from 34 storms.

hardwoods. Six treatments, which included the most common methods to timber harvesting and site preparation, were chosen to study their effects on average soil water content of the whole soil profile (0–1.35 m) (1) undisturbed forest with full crown closure (0.29 g cm^{-3}); (2) thinned forest, 50% of its original crown density (0.40 g cm^{-3}; (3) clearcut, all merchantable timber removed, no site preparation (0.41 g cm^{-3}); (4) clearcut and roller-chopped (0.44 g cm^{-3}); (5) clearcut, sheared, root-raked and slash-piled in windrows (0.43 g cm^{-3}); and (6) clearcut, clean-tilled, continuous fallow, cultivated up and down hill (0.49 g cm^{-3}).

The paper provides a prodigious amount of information and is rewarding and clear to read.

A similar, but earlier, study was made in the central Appalachian mountains in West Virginia, USA (Patric 1973). The forest was deciduous with leaves emerging in April, being fully grown in June and falling in early October. Rainfall was 1450 mm yr^{-1}, soil depth ranged from 55–120 cm in silty clay loam, and neutron probe access tubes were 50 cm deep. There were two watersheds, forested and barren (vegetation treated with herbicide), with 25 access tubes in each watershed. Substantially more water was found in the barren catchment during the growing season, but after autumn both catchments had similar soil water regimes.

Eastham and Rose (1988) in north-eastern Australia (see Sect. 4.2) used zero, medium and high tree density treatments and lysimeters to determine soil water content over 2 years. Their data showed that soil water was consistently highest under the medium density treatment. On that treatment, the tree canopy produced sufficient shade to reduce the high evaporation which caused low soil water content at low tree density. Further, the incomplete canopy at the medium tree density allowed greater throughfall than occurred under the closed canopy at the high tree density. Because of the clever experimental design and precise data (except during frost damage) on soil water and evaporation obtained from the lysimeters, clear relationships between these components of water balance were established. This implies that a simple model would increase the applicability of the results.

Although I have developed a separate section for the storage of soil water, it is really part of the hydrologic recharge system of the permanent, unconfined aquifer (water table). Yet, we have seen in Sect. 2.3.2 that its importance for recharge depends on the conducting systems in the soil. High conductivity systems bypass the soil storage. It is not often that water storage researchers investigate the presence and evaluate the flow characteristics of such systems. They may be far more prevalent than expected. The research is easy enough to operate. It is really a problem of unawareness, particularly so for agriculturalists for whom groundwater recharge has been beyond their mandate.

Whether or not there is a dual or multiple conductivity system operating, the bulk water storage still influences recharge though perhaps orders of magnitude more slowly.

Another soil water bypass system is the perched, or temporary water table near the soil surface. I have neglected it. Since agricultural practices tend to develop compaction layers, it might be expected that forested soils develop better vertical conduction systems apart from any other modification caused by forest litter to the soil water conduction system.

4.5 Groundwater Recharge and Discharge

The effect of deforestation on the groundwater is profound, and the "untidiness" below ground makes it complex. Section 2.5.1 gave a brief statement of the main issues.

A major collection of information on forest clearing and groundwater in south–west Australia can be found in the special issue of the *Journal of Hydrology* 94, 1987. I shall begin there.

Peck and Williamson (1987) compared groundwater levels in paired lateritic catchments which were cleared or left intact. The climate is Mediterranean and they chose catchments in the 1200 and 800 mm yr^{-1} rainfall zones (no ice or snow). The jarrah forest (*Eucalyptus marginata*) grows on a gravelly, permeable upper layer below which is a sandy kaolinitic clay in which the unconfined aquifer is located. The lateritic element does not exist in the lower parts of catchments. Water tables are 5–20 m deep upslope. In the lower parts, a siliceous confining layer keeps the aquifer under pressure through which it may discharge to the surface through leaks. Such aquifers are permanent over the entire catchment in the high rainfall zone but are permanent only in the lower parts of catchments in the lower rainfall zone. Hydraulic conductivity varies by more than a factor of 100, which suggests the existence of preferred pathways.

Johnston (1987a, b) studied this phenomenon in the high rainfall zone. Although aware of the difficulties with using a one-dimensional steady-state model of water and chloride, he considered it to be an excellent tool for identifying features of subsurface hydrology. He investigated vertical water flux densities over a 2 year period with particular care. There were two populations in the unsaturated zone.

Over most of the area, below 5 m, flux density was 2–7 mm yr^{-1}. Over the remaining small area the values ranged from 50–100 mm yr^{-1}, a value which closely matched the apparent rate of recharge to the groundwater. A groundwater mound was seen to develop below it some 12 h after intense storms and then to dissipate over 2–4 days. Johnston's work is a leading study of below-ground complexity at depth.

Returning to Peck and Williamson, in the higher rainfall catchments, the potentiometric surface under forest responded by 1–4 m after each winter wet season, indicating significant seasonal recharge which was widespread. By contrast, the seasonal response was rarely more than 1 m under forest in the medium rainfall zone, and in many sites there was no apparent recharge in several years of study.

In areas fully cleared for agriculture, the potentiometric surface moved up by more than 2.6 m yr^{-1} averaged over several years (6–12% of rainfall). Under partial clearing, the potentiometric surface rose less rapidly (0.9 m yr^{-1}).

Hookey (1987) developed the above research further modelling groundwater levels on a whole catchment basis. The model also predicted the delays in ground-water response to forest clearing.

Sharma and Craig (1989)used the depth distribution of natural chloride to estimate recharge under native woodland and pine plantations on a groundwater mound supplying water to a city. The climate is Mediterranean with a rainfall of 800 mm yr^{-1}. Although, the soil is a deep (> 30 m) and apparently uniform sand, they encountered perferred pathways!

Natural chloride estimates of recharge require a steady state, which is valid for the woodland (100 years) but not for the 25-year- and 8-year-old pine plantations. Making due allowance for that, they estimated recharge to be 120 mm yr^{-1} under woodland, > 4 mm yr^{-1} under mature pines and 245 mm yr^{-1} under the young pines. The implication for water supply is clear. The perferred pathways under woodland contributed over 50% of recharge, but under the young pines such an effect was marginal. Why?

The effect of vegetation treatment on groundwater recharge was shown by Heikurainen (1967) to be modified by surface drainage in northern Europe's peatlands. Clearcutting and thinning both increase recharge, but because the water table is near the surface (> 0.5 m), much of the runoff discharges into the drains. The cause of the increased recharge with thinning is due to the small values of total evaporation whence an increase in throughfall. Heikurainen does not mention the effect of thinning on snowmelt and consequences for recharge.

In those climatic regions where deforestation results in greatly increased throughfall, there will also be increases in surface and subsusrface runoff, recharge to the water table, and discharge into streams. Water moves at different velocities in each of those pathways, all of which can be measured. Leakage out of the catchment which bypasses the stream, or leakage into the catchment from aquifer systems below are usually undetected, unmeasured, and perhaps unmeasurable even if identified.

The references in Sects. 4.4 and 4.5 have impinged on the "Russian paradox". I do not think that they have helped to explain it. The Russian paradox may be resolved only when we have developed the ability to measure water balance parameters accurately on a macro-scale. Lawson (1986) takes us in that direction in his highly readable account of the effects of deforestation and induced changes in a tropical meso/micro-climate.

5 The Case for Optimism

Klemeš (1988) summarized three important reasons why the present era offers extraordinary opportunities: (1) the growth of the scale of human interference with nature; (2) the advances in observational and data handling technologies; and (3) the growing collaboration and interdependence between biogeophysical sciences as they tackle problems on continental and global scales. I would like to add a fourth reason: the growing contribution from socio-economic research is helping funding bodies and decision makers to become aware of hydrologic issues and projects which need finance for them to be implemented.

The greatest cause for optimism is the current high quality of ideas, rationalization and implementation now available or developing. The critical advances which are being made in technology will help to resolve many current difficulties in quantification, and therefore understanding of hydrologic, vegetation and climatic processes. Parallel advances in modelling are assisting to make the most of technical capability by extending its application and our understanding of large, complex, interactive systems.

Section 2 of this review reveals that recent research in each component of the water balance equation has advanced, significantly, the accuracy, scale and practicability of measurement. Traditionally, such achievements have been considered worthy ends in themselves. New techniques and models are stimulants, and their creators attain prestige. Yet, collectively, we have failed. For those new techniques, while giving credibility to process research, have another role to play which we have tended to ignore. The resolution of major regional and global management issues, such as deforestation, requires those techniques to be applied to immense scales. Accordingly, creators or developers of new procedures ought to go further than to publish their achievements in a journal. They should become the entrepreneurs of rigorous, widespread application.

In surveying the history of water balance research over three decades, I did notice a minor surge in the application of new methods during the second half of the 1980s. And those papers tended to have multidisciplinary authorships. Such trends will lead to more effective research in the management of major problems with vegetation and water balance.

Multidisciplinary projects need appropriate media for their comprehensive reporting and reviewing. The establishment of Advances in Bioclimatology is therefore timely and appropriate. And, if our optimism is well founded, it will also become important to have a similarly titled series for original multidisciplinary research and modelling papers in the broad field of deforestation, revegetation, water balance and climate.

Throughout the writing of this review, my optimism was sustained by the paper of Edwards and Blackie (1981), reported in Sect. 3. They found that forests can, with careful management, be converted to specific forms of "agriculture" without substantial long-term change in the components of water balance and erosion.

The general challenge for hydrologists in devising deforestation and revegetation programs is to specify new vegetation regimes which will not seriously perturb water balance and soil stability. It is the prior responsibility of citizens and of governments to decide whether that deforestation should be permitted.

May the readers of this review share my optimism and may future research justify my enthusiasm!

Acknowledgments. I have been skillfully and warmly assisted in research and preparation of this review by Bernadette Waugh, who undertook the electronic reference search; by Linley Thornber, who converted electrons into hard copy, and by Helen Warrener who, magically it seemed to me, turned my notes into a professional manuscript. Lawrence Hamilton of the East–West Center, Hawaii, provided me with two of his annotated bibliographies which helped me through the earlier literature. Gerald Stanhill encouraged me at crucial stages and confronted me with "the Russian paradox". The "anonymous" referees helped me with some short comings and the editorial staff at Springer made further improvments. Thank you all for turning a daunting assignment into one much less so!

References

Abdulmumin S, Myrup LO, Hatfield JL (1987) An energy balance approach to determine regional evapotranspiration based on planetary boundary layer similarity theory and regularly recorded data. Water Resour Res 23:2050–2058

Allaway J, Cox PMJ (1989) Forests and competing land uses in Kenya. Environ Manage 13:171–187

Allerup P, Madsen H (1980) Accuracy of point precipitation measurements. Nord Hydrol 11:57–70

Alley WM (1984) On the treatment of evapotranspiration, soil moisture accounting, and aquifer recharge in monthly water balance models. Water Resour Res 20:1137–1149

Anyadike RNC (1987) The Linacre evaporation formula tested and compared to others in various climates over West Africa. Agric For Meteorol 39:111–119

Aston AR, Dunin FX (1980) Land use hydrology: Shoalhaven, New South Wales. J Hydrol 48:71–87

Athavale RN, Murti CS, Chand R (1980) Estimation of recharge to the phreatic aquifers of the Lower Maner Basin, India, by using the tritium injection method. J Hydrol 45:185–202

Ayoade JO (1976) Evaporation and evapotranspiration in Nigeria. J Trop Geogr 43:9–19

Baker MB Jr (1986) Effects of ponderosa pine treatments on water yield in Arizona. Water Resour Res 22:67–73

Barber C, Briegel D (1987) A method for the in situ determination of dissolved methane in groundwater in shallow aquifers. J Contam Hydrol 2:51–60

Barnes CJ, Allison GB (1988) Tracing of water movement in the unsaturated zone using stable isotopes of hydrogen and oxygen. J Hydrol 100:143–177

Berris SN, Harr RD (1987) Comparative snow accumulation and melt during rainfall in forested and clearcut plots in the Western Cascades of Oregon. Water Resour Res 23:135–142

Bethlahmy N (1974) More stream flow after a bark beetle epidemic. J Hydrol 23:185–189

Bevan K (1989) Changing ideas in hydrology – the case of physically based models. J Hydrol 105:157–172

Biddiscombe EF, Rogers AL, Greenwood EAN, DeBoer ES (1981) Establishment and early growth of species in farm plantations near salt seeps. Aust J Ecol 6:383–389

Black PE (1968) Stream flow increases following abandonment on eastern New York watershed. Water Resour Res 4:1171–1178

Blackburn WH, Wood JC, DeHaven MG (1986) Storm flow and sediment losses from site-prepared forestland in east Texas. Water Resour Res 22:776–784

Blackie JR (1987) The Balquhidder catchments, Scotland: the first four years. Trans R Soc Edinb Earth Sci 78:227–239

Blackie JR, Newson MD (1986) The effects of forestry on the quantity and quality of run-off in upland Britain. In: de LG Solbe JF (ed) Effects of land use in fresh waters. Ellis Horwood, Chichester, 568 pp

Bonell M, Gilmour DA, Sinclair DF (1981) Soil hydraulic properties and their effect on surface and subsurface water transfer in tropical rainforest catchment. Hydrol Sci Bull 26:1–18

Bonell M, Cassells DS, Gilmour DA (1982) Vertical and lateral soil water movement in a tropical rainforest catchment. In: O'Loughlin EM, Bren LJ (eds) The first national symposium on forest hydrology. Institution of Engineers Aust Nat Conf Publ No 82/6, pp 30–38

Borg H, Stoneman GL, Ward CG (1987) Stream and groundwater response logging and subsequent regeneration in the southern forest of Western Australia. Tech Rept No 16, Dept of Conservation and Land Management, Como, West Australia, pp 1–113

Bosch JM, Hewlett JD (1982) A review of catchment experiments to determine the effects of vegetation changes on water yield and evapotranspiration. J Hydrol 55:3–23

Bosman HH (1987) The influence of installation practices on evaporation from Symon's tank and American Class A-pan evaporimeters. Agric For Meteorol 41:307–323

Bouchet RJ (1963) Evapotranspiration reele et potentielle, signification climatique, Publ 62. General Assembly Berkeley, Int Assoc Sci Hydrol Gentbrugge, Belgium, pp 134–142

Bren LJ, Leitch CJ (1985) Hydrologic effects of a stretch of forest road. Aust For Res 15:183–194

Bren LJ, Leitch CJ (1986) Rainfall and water yields of three small, forested catchments in north–east Victoria, and relation to flow of local rivers. Proc R Soc Vict 98:19–29

Brooks KN (1985) Evaluation of deforestation impacts on environment and productivity. Proc IXth World Forestry Congr, Mexico, E-1.6.1A, 13 pp

Brown JAH (1972) Hydrologic effects of a bushfire in a catchment in south-eastern New South Wales. J Hydrol 15:77–96

Brown TC, O'Connell PF, Hibbert AR (1974) Chaparral conversion potential in Arizona. Part II. An economic analysis. USDA For Serv RM-127, Rocky Mt For Range Exp Stn Fort Collins, Colorado, 28 pp

Bruijnzeel LA (1983) Evaluation of runoff sources in a forested basin in a wet monsoonal environment: a combined hydrological and hydrochemical approach. In: Hydrology of humid tropical regions. IAHS Publ No 140, pp 165–174

Bruijnzeel LA (1986) Environmental impacts of (de) forestation in the humid tropics. A watershed perspective. Wallaceana 46:3–13

Bruijnzeel LA (1990) Hydrology of moist tropical forests and effects on conversion: a state of knowledge review. UNESCO Int Hydrol Programme, Humid Tropics Programme, Free University, Amsterdam, pp 1–224

Burch GJ, Bath RK, Moore ID, O'Loughlin EM (1987) Comparative hydrological behaviour of forested and cleared catchments in south-eastern Australia. J Hydrol 90:19–42

Burch GJ, Moore ID, Burns J (1989) Soil hydrophobic effects on infiltration and catchment runoff. Hydrol Processes 3:211–222

Burn CR, Smith MW (1985) Comment on "Water movement into seasonally frozen soils" by DL Kane and J Stein. Water Resour Res 21:1051–1052

Burt TP, Williams PJ (1976) Hydraulic conductivity in frozen soils. Earth Surf Processes 1:349–360

Buselli G, Barber C, Williamson DR (1986) The mapping of groundwater contamination of soil salinity by electromagnetic methods. In: Inst Engineers Australia Hydrology and Water Resources Symposium, Griffith University, Brisbane, pp 317–322

Buttle JM (1989) Soil moisture and groundwater responses to snowmelt on a drumlin sideslope. J Hydrol 105:335–355

Buttle JM, McDonnell JJ (1987) Modelling the areal depletion of snowcover in a forested catchment. J Hydrol 90:43–60

Byrne GF, Dunin FX, Diggle PJ (1988) Forest evaporation and meteorological data: a test of a complementary theory advection – aridity approach. Water Resour Res 24:30–34

Calder IR (1985) What are the limits on forest evaporation? A comment. J Hydrol 82:179–192

Calder IR (1986a) The influence of land use on water yield in upland areas of the UK. J Hydrol 88:201–211

Calder IR (1986b) Water use of eucalypts – a review with special reference to South India. Agric Water Manage 11:333–342

Calder IR (1986c) A stochastic model of rainfall interception. J Hydrol 89:65–71

Calder IR (1990) Evaporation in the uplands. John Wiley, Chichester, 148 pp

Calder IR, Newson MD (1979) Land-use and upland water resources in Britain – a strategic look. Water Resour Bull 15:1628–1639

Calder IR, Wright IR, Murdiyarso D (1986) A study of evaporation from tropical rain forest – West Java. J Hydrol 89:13–31

Caprio JM, Grunwald GK, Snyder RD (1989a) Conservation and storage of snowmelt in stubble land and fallow under alternate fallow-strip cropping management in Montana. Agric For Meteorol 45:265–279

Caprio JM, Grunwald GK, Snyder RD (1989b) Effects of climate on potential soil water gain from snowmelt in stubble and fallow fields. Agric For Meteorol 45:281–298

Carroll SS, Carroll TR (1989) Effect of uneven snow cover on airborne snow water equivalent estimates obtained by measuring terrestrial gamma radiation. Water Resour Res 25:1505–1510

Čermák J, Jeník J, Kučera J, Žídec V (1984) Xylem water flow in a crack willow tree (*Salix fragilis* L.) in relation to diurnal changes of environment. Oecologia 64:145–151

Chang M, Watters SP (1984) Forests and other factors associated with streamflows in East Texas. Water Resour Bull 20:713–719

Chang M, Ting JC, Wong KL, Hunt EV Jr (1983) Soil moisture regimes as affected by silvicultural treatments in humid East Texas. IAHS Publ No 140, pp 175–186

Cohen Y, Fuchs M, Green GC (1981) Improving the heat pulse method for determination of sap flow in trees. Plant Cell Environ 4:391–397

Cohen Y, Kelliher FM, Black TA (1985) Determination of sap flow in Douglas-fir trees using the heat pulse technique. Can J For Res 15:422–428

Collier CG (1986a) Accuracy of rainfall estimates by radar. Part I. Calibration by telemetering raingauges. J Hydrol 83:207–223

Collier CG (1986b) Accuracy of rainfall estimates by radar. Part II. Comparison with raingauge network. J Hydrol 83:225–235

Collier CG (1986c) Accuracy of rainfall estimates by radar. Part III. Application for short-term flood forecasting. J Hydrol 83:237–249

Constantz J, Herkelrath WN, Murphy F (1988) Air encapsulation during infiltration. Soil Sci Soc Am J 52:10–16

Cooper TA, Lockwood JG (1987) The influence of rainfall distribution in numerical simulation of evapotranspiration from a multilayer model pine canopy. Water Resour Res 23:1645–1656

Cordery I, Pilgrim DH (1983) On the lack of dependence of losses from flood runoff on soil and cover characteristics. In: Hydrology of humid tropical regions. IAHS Publ No 140, pp 187–195

Costin AB (1967) Management opportunities in Australian high mountain catchments. In: Sopper WE, Lull HW (eds) Forest hydrology. Pergamon, Oxford, pp 565–577

Crockford RH, Richardson D (1990) Partitioning of rainfall in a eucalypt forest and pine plantation in south-eastern Australia. I. Throughfall measurement in a eucalypt forest: effect of method and species composition. II. Stemflow and factors affecting stemflow in a dry sclerophyll eucalypt forest and a *Pinus radiata* plantation. III. Determination of canopy storage capacity of a dry sclerophyll eucalypt forest. IV. The relationship of interception and canopy storage capacity, the interception of these forests, and the effect on interception of thinning the pine plantation. Hydrol Processes 4:131–188

De Bruin HAR (1983) Evapotranspiration in humid tropical regions. In: Hydrology of humid tropical regions. IAHS Publ No 140, pp 299–311

De Walle DR, Lynch JA (1975) Partial forest clearing effects on snowmelt runoff. In: Watershed management. American Society of Civil Engineers, New York, pp 337–346

Dickinson RE (1989) Modeling the effects of Amazonian deforestation on regional surface climate: a review, Agric For Meteorol 47:349–347

Dolman AJ (1987) Summer and winter rainfall interception in an oak forest. Predictions with an analytical and a numerical simulation model. J Hydrol 90:1–9

Dolman AJ (1988) Transpiration of an oak forest as predicted from porometer and weather data. J Hydrol 97:225–234

Dolman AJ, Stewart JB, Cooper JD (1988) Predicting forest transpiration form climatological data. Agric For Meteorol 42:339–353

Dons A (1986) The effect of large-scale afforestation on Tarawera River flows. J Hydrol NZ 25:61–73

Douglass JE (1983) The potential for water yield augmentation from forest management in the eastern United States. Water Resour Bull 19:351–358

Dugas WA, Bland WL (1989) The accuracy of evaporation measurements from small lysimeters. Agric For Meteorol 46:119–129

Dunin FX (1991) Extrapolation of "point" measurements of evaporation: some issues of scale. In: Henderson-Sellers A, Pitman AJ (eds) Vegetation and climate interactions in semi-arid regions. Vegetatio 91:39–47. Kluwer, Academic Publishers

Dunin FX, Greenwood EAN (1986) Evaluation of the ventilated chamber for measuring evaporation from a forest. Hydrol Processes 1:47–62

Dunin FX, O'Loughlin EM, Reyenga W (1988) Interception loss from eucalypt forest: lysimeter determination of hourly rates for long term evaluation. Hydrol Processes 2:315–329

Dunin FX, Nulsen RA, Baxter IN, Greenwood EAN (1989) Evaporation from a lupin crop – a comparison of methods. Agric For Meteorol 46:297–311

Dunne T, Leopold LB (1978) Water in environmental planning. Freeman, San Francisco, 818 pp

Eastham J, Rose CW (1988) Pasture evapotranspiration under varying tree planting density in an agroforestry experiment. Agric Water Manage 15:87–105

Eastham J, Rose CW, Cameron DM, Rance SJ, Talsma T (1988) The effect of tree spacing on evaporation from an agroforestry experiment. Agric For Meteorol 42:355–368

Edwards KA, Blackie JR (1981) East African catchment experiments 1958–1974. In: Lal R, Russell EW (eds) Tropical agricultural hydrology. Wiley & Sons, London, pp 163–188

Edwards WRN, Warwick NWM (1984) Transpiration from a kiwifruit vine as estimated by the heat pulse technique and the Penman–Monteith equation. NZ J Agric Res 27:537–543

Engel R, McFarlane DJ, Street G (1987) The influence of dolerite dykes on saline seeps in south-western Australia. Aust J Soil Res 25:125–136

Farrington P, Greenwood EAN, Bartle GA, Beresford JD, Watson GD (1989) Evaporation from a Banksia woodland on a groundwater mound. J Hydrol 105:173–186

Ffolliott PF, Gottfried GJ, Baker Jr MB (1989) Water yield from forest snowpack management: research findings in Arizona and New Mexico. Water Resour Res 25:1999–2007

Fichtner K, Schulze E-D (1990) Xylem water flow in tropical vines as measured by a steady state heating method. Oecologia 82:355–361

Freeze RA (1972a) Role of subsurface flow in generating surface runoff. 1. Base flow contributions to channel flow. Water Resour Res 8:609–623

Freeze RA (1972b) Role of subsurface flow in generating surface runoff. 2. Upstream source areas. Water Resour Res 8:1272–1283

Galbraith AF (1975) Method for predicting increases in water yield related to timber harvesting and site conditions. In: Watershed management. American Society Civil Engineers, New York, pp 169–184

Garbrecht J, Shen HW (1988) The physical framework of the dependence between channel flow hydrographs and drainage network morphometry. Hydrol Processes 2:337–355

Garczynski F (1980) Effect of percentage forest cover on the hydrological regime in three regions of the USA. IAHS-AISH Publ No 130, pp 67–74

Gash JHC (1979) An analytical model of rainfall interception by forests. QJR Meteorol Soc 105:43–55

Gash JHC, Wright IR, Lloyd CR (1980) Comparative estimates of interception loss from three coniferous forests in Great Britain. J Hydrol 48:89–105

Gat JR, Matsui E, Salati E (1985) The effect of deforestation on the water cycle in the Amazon Basin: an attempt to reformulate the problem. Acta Amazonica 15:307–310

German PF, Douglas LA (1987) Comments on "Particle transport through porous media" by McDowell-Boyer LM, Hunt J, Sitar N. Water Resour Res 23:1697–1698

Gilmour DA (1977) Effect of rainforest logging and clearing on water yield and quality in a high rainfall zone of north–east Queensland. Proc Symp Hydrology of Northern Australia. The Institution of Engineers, Australia, pp 156–160

Gilmour DA, Bonell M, Cassells DS (1987) The effects of forestation on soil hydraulic properties in the middle hills of Nepal. A preliminary assessment. Mountain Res Dev 7:239–249

Golding DL, Swanson RH (1986) Snow distribution patterns in clearings and adjacent forest. Water Resour Res 22: 1931–1940

Granger RJ (1989a) An examination of the concept of potential evaporation. J Hydrol 111:9–19

Granger RJ (1989b) A complementary relationship approach for evaporation from nonsaturated surfaces. J Hydrol 111:31–38

Granger RJ, Gray DM (1989) Evaporation from natural nonsaturated surfaces. J Hydrol 111:21–29

Greacen EL, Sands R (1980) Compaction of forest soils. Aust J Soil Res 18:163–189

Greenwood EAN (1986) Water use by trees and shrubs for lowering saline groundwater. Reclam Reveg Res 5:423–434

Greenwood EAN (1988) The hydrologic role of vegetation in the development and reclamation of dryland salinity. In: Allen EB (ed) The reconstruction of disturbed arid lands – an ecological approach. AAAS Sel Symp 109:205–233

Greenwood EAN, Beresford JD, Bartle JR (1981) Evaporation from vegetation in landscapes developing secondary salinity using the ventilated chamber technique. III. Evaporation from a *Pinus radiata* tree and surrounding pasture in an agroforestry plantation. J Hydrol 50:155–166

Greenwood EAN, Klein L, Beresford JD, Watson GD (1985a) Differences in annual evaporation between grazed pasture and *Eucalyptus* species in plantations on a saline farm catchment. J Hydrol 78:261–278

Greenwood EAN, Klein L, Beresford JD, Watson GD, Wright KD (1985b) Evaporation from the understorey in the jarrah (*Eucalyptus marginata* Don ex Sm.) forest, southwestern Australia. J Hydrol 80:337–349

Gross R, Eeles CWO, Gustard A (1989) Application of a lumped conceptual model to FREND catchments. IAHS Publ 187:309–320

Haltiner JP, Salas JD (1988) Short-term forecasting of snowmelt runoff using ARMAX models. Water Resour Bull 24:1083–1089

Hamilton LS (1987) What are the impacts of Himalayan deforestation on the Ganges–Brahmaputra lowlands and delta? Assumptions and facts. Mountain Res Dev 7:256–263

Harding RJ (1986) Exchanges of energy and mass associated with a melting snowpack. In: Modelling snowmelt induced processes. IAHS Publ No 155, pp 3–15

Harr RD (1982) Fog drip in the Bull Run municipal watershed, Oregon. Water Resour Bull 18:785–789

Harr RD (1986) Effects of clear cutting on rain-on-snow runoff in western Oregon: A new look at old studies. Water Resour Res 22:1095–1100

Harr RD, McCorison FM (1979) Initial effects of clearcut logging on size and timing of peak flows in a small watershed in westren Oregon. Water Resour Res 15:90–94

Harr RD, Harper WC, Krygier JT, Hsieh FS (1975) Changes in storm hydrographs after road building and clear-cutting in the Oregon coast range. Water Resour Res 11:436–444

Harris AR (1972) Infiltration rate as affected by soil freezing under three cover types. Soil Sci Soc Am Proc 36:489–492

Hatton TJ, Vertessey RA (1990) Transpiration of plantation *Pinus radiata* estimated by the heat pulse method and the Bowen ratio. Hydrol Processes 4:289–298

Heikurainen L (1967) Effect of cutting on the ground-water level on drained peatlands. In: Sopper WE, Lull HW (eds) Forest hydrology Pergamon, Oxford, pp 345–354

Helvey JD (1980) Effects of a northcentral Washington wildfire on runoff and sediment production. Water Resour Bull 16:627–634

Hetrick DM, Travis CC, Shirley PS, Etnier EL(1986) Model predictions of watershed hydrologic components: comparison and verification. Water Resour Bull 22:803–810

Hewlett JD, Bosch JM (1984) The dependence of storm flows on rainfall intensity and vegetal cover in South Africa J Hydrol 75:365–381

Hewlett JD, Helvey JD (1970) Effects of forest clear felling on the storm hydrograph. Water Resour Res 6:768–782

Hewlett JD, Troendle CA (1975) Non-point and diffused water sources: a variable source area problem. In: Watershed management American Society of Civil Engineers, New York, pp 21–46

Hewlett JD, Fortson JC, Cunningham GB (1977) Effect of rainfall intensity on storm flow and peak discharge from forest land. Water Resour Res 13:259–266

Hibbert AR (1967) Forest treatment effects on water yield. In: Sopper WE, Lull HW (eds) Forest hydrology. Pergamon, Oxford, pp 527–543

Hibbert AR (1969) Water yield changes after converting a forested catchment to grass. Water Resour Res 5:634–640

Hibbert AR (1983) Water yield improvement potential by vegetation management on western rangelands. Water Resour Bull 19:375-381

Hibbert AR, Davis EA, Brown TC (1975) Managing chaparral for water and other resources in Arizona. In: Watershed management American Society of Civil Engineers, New York, pp 445-467

Hoggan DH, Peters JC, Loehlein W (1987) Real-time snow simulation model for the Monongahela River basin. Water Resour Bull 23:1141-1147

Hookey GR (1987) Prediction of delays in groundwater response to catchment clearing. J Hydrol 94:181-198

Hornbeck JW (1975) Stream flow response to forest cutting and revegetation. Water Resour Bull 11:1257-1260

Hornbeck JW, Pierce RS, Federer CA (1970) Stream flow changes after forest clearing in New England. Water Resour Res 6:1124-1132

Hoshi T, Uchida S, Kotoda K (1989) Development of a system to estimate evapotranspiration over complex terrain using Landsat MSS, elevation and meteorological data. Hydrol Sci J 34:635-649

Hsia YJ, Koh CC (1983) Water yield resulting from clearcutting a small hardwood basin in central Taiwan. In: Hydrology of humid tropical regions with particular reference to the hydrological effects of agriculture and forestry practice. IAHS Publ No 140, pp 215-220

Ingraham NL, Matthews RA (1988) Fog drip as a source of groundwater recharge in northern Kenya. Water Resour Res 24:1406-1410

Jackson IJ (1978) Local differences in patterns of variability of tropical rainfall: some characteristics and implications. J Hydrol 38:273-287

Jarvis PG, McNaughton KG (1986) Stomatal control of transpiration: scaling up from leaf to region. Adv Ecol Res 15:1-49

Johnston CD (1987a) Distribution of environmental chloride in relation to subsurface hydrology. J Hydrol 94:67-88

Johnston DC (1987b) Preferred water flow and localised recharge in a variable regolith. J Hydrol 94:129-142

Johnston CD, Hurle HD, Hudson DR, Height MI (1983) Water movement through preferred paths in lateritic profiles of the Darling Plateau Western Australia. CSIRO Division of Groundwater Research Tech Pap No 1 Perth, Western Australia, 34 pp

Jupp DLB, Kalma J (1990) Distributing evapotranspiration in a catchment using airbourne remote sensing. Asian pac Remote Sens J 2:13-26

Jupp DLB, Walker J, Kalma J, Smith R (1990) Remote sensing of change in components of the regional water balance of the Murray-Darling Basin using satellite imaged and spatially registered environmental data. Aust J Soil Res 28:409-15

Kachanoski RG, deJong E (1988) Scale dependence and the temporal persistence of spatial patterns of soil water storage. Water Resour Res 24:85-91

Kaluarachchi JJ, Parker JC (1987) Effects of hysteresis with air entrapment on water flow in the unsaturated zone. Water Resour Res 23:1967-1976

Kane DL, Stein J (1983) Water movement into seasonally frozen soils. Water Resour Res 19:1547-1557

Kattelmann RC, Berg NH, Pack MK (1985) Estimating regional snow water equivalent with a simple simulation model. Water Resour Bull 21:273-280

Kay JA (1984) Discussion "Sampling accuracy of pit vs standard rain gages on the Fernow experimental forest" by JD Helvey and JH Patric. Water Resour Bull 20:275-276

Kelliher FM, Black TA, Price DT (1986) Estimating the effects of understorey removal from a Douglas fir forest using a two-layer canopy evapotranspiration model. Water Resour Res 22:1891-1899

Kerfoot O (1968) Mist precipitation on vegetation. For Abstr 29:8-20

Klemeš V (1986) Dilettantism in hydrology: transition or destiny? Water Resour Res 22:177s-188s

Klemeš V (1988) A hydrological perspective. J Hydrol 100:3-28

Kramer PJ (1988) Changing concepts regarding plant water relations. Plant Cell Environ 11:565-568

Krutilla JV, Bowes MD, Sherman P (1983) Watershed management for joint production of water and timber: a provisional assessment. Water Resour Bull 19:403-414

Kurz WA, Kimmins JP (1987) Analysis of some sources of error in methods used to determine fine root production in forest ecosystems: a simulation approach. Can J For Res 17:909-912

Lal R (1981) Deforestation of tropical rainforest and hydrological problems. In: Lal R, Russell EW (eds) Tropical agricultural hydrology. John Wiley, Chichester

Lal R, Cummings DJ (1979) Clearing a tropical forest. I. Effects on soil and micro climate. Field Crops Res 2:91–107

Langford KJ (1976) Changes in yield of water following a bushfire in a forest of *Eucalyptus regnans*. J Hydrol 29:87–114

Langford KJ, Moran RJ O'Shaugnessy PJ (1980) The North Maroondah experiment pretreatment phase comparison of catchment water balances. J Hydrol 46:123–145

Lawson TL (1986) Deforestation and induced changes in meso/microclimate. In: Lal R, Sanchez PA, Cummings RW Jr (eds) Land clearing and development in the tropics. A.A. Balkema, Rotterdam, pp 195–202

Leavesley GH (1989) Problems of snowmelt runoff modelling for a variety of physiographic and climatic conditions. Hydrol Sci J 34:617–634

Lebel T, Bastin G, Obled C, Creutin JD (1987) On the accuracy of areal rainfall estimation: a case study. Water Resour Res 23:2123–2134

Leith RM, Solomon SI (1985) Estimation of precipitation, evapotranspiration and runoff using GOES. In: Advances in evapotranspiration. American Society of Agricultural Engineers, St. Joseph MI, pp 366–376

Leopoldo PR, Franken W, Salati E, Ribeiro MN (1987) Towards a water balance in the Central Amazonian region. Experientia 43:222–233

Lettau H, Lettau K, Molion LCB (1979) Amazonia's hydrologic cycle and the role of atmospheric recycling in assessing deforestation effects. Monthly Weather Rev 107:227–238

Linacre ET (1977) A simple formula for estimating evaporation rates in various climates using temperature data alone. Agric Meteorol 18:409–424

Liu Y (1987) The influence of variation in forest cover on design floods. J Hydrol 96:367–374

Lloyd CR, de O Marques FA (1988) Spatial variability and throughfall and stemflow measurements in Amazonian rainforest. Agric For Meteorol 42:63–73

Lloyd CR, Gash JHC, Shuttleworth WJ, de O Marques FA (1988) The measurement and modelling of rainfall interception loss from Amazonian rainforest. Agric For Meteorol 43:277–294

Lusby GC (1979) Effects of converting sagebrush cover to grass on the hydrology of small watersheds at Boco Mountain, Colorado. Geol Surv Water Supply Pap 1532J:1–24

Luvall JC, Parker G, Jordan C (1985) Tropical deforestation and evapotranspiration. In: Quinones F, Sanchez AV (eds) Proc Int Symp Tropical Hydrology and 2nd Caribbean Islands Water Resources Congress, San Juan, Puerto Rico, Am Water Resources Assoc, pp 7–10

McDonald MG, Harbaugh AW (1984) A modular three-dimensional finite-difference groundwater flow model. US Geol Surv Open-File Rep 83–875, 528 pp

McDowell-Boyer LM, Hunt JR, Sitar N (1986) Particle transport through porous media. Water Resour Res 22:1901–1921

McGuinness JL, Harrold LL (1971) Reforestation influences on small watershed stream flow. Water Resour Res 7:845–852

McIlroy IC, Dunin FX (1982) A forest evaporation technique comparison experiment. 1st Nat Symp Forest Hydrology, Aust Nat Conf Publ No 82/6. The Institution of Engineers, Australia pp 12–17

McNaughton KG, Spriggs TW (1989) An evaluation of the Priestley and Taylor equation and the complementary relationship using the results from a mixed-layer model of the convective boundary layer. In: Black TA, Spittlehouse DL, Novak MD, Price DT (eds) Estimation of areal evapotranspiration. IAHS Publ No 177, pp 89–104

Morris EM (1989) Turbulent transfer over snow and ice. J Hydrol 105:205–223

Morton FI (1983) Operational estimates of areal evapotranspiration and their significance to the science and practice of hydrology. J Hydrol 66:1–76

Morton FI (1984) What are the limits on forest evaporation? J Hyrol 74:373–398

Morton FI (1985) The complementary relationship areal evapotranspiration model: How it works. In: Advances in evapotranspiration. American Society of Agricultural Engineers, pp 377–384

Mulder JPM (1985) Simulating interception loss using standard meteorological data. In: Hutchinson BA, Hicks BB (eds) The forest-atmosphere interaction. D Reidel, Dordrecht, pp 177–196

Mumeka A (1986) Effect of deforestation and subsistence agriculture on runoff of the Kafue River, headwaters, Zambia. Hydrol Sci 31:543–554

Nakano Hi (1967) Effects of changes of forest conditions on water yield, peak flow and direct runoff of small watersheds in Japan. In: Sopper WE, Lull HW (eds) Forest hydrology. Pergamon, Oxford, pp 551–564

Nash JE (1989) Potential evaporation and "the complementary relationship" J Hydrol 111:1–7

Obeysekera JTB, Tabios GQ III, Salas JD (1987) On parameter estimation of temporal rainfall models. Water Resour Res 23:1837–1850

O'Connell MJ, O'Shaughnessy PJ (1975) The Wallaby Creek fog drip study. Melbourne & Metropolitan Board of Works Rep No MMBW-W-0004, Melbourne, Asutralia

Oliver HR (1983) The availablity of evaporation data in space and time for use in water balance computations. In: Van der Beken A, Herrmann A (eds) New approaches to water balance computations. IAHS Publ No 148 pp 21–31

Parker GG, Luval JC, Jordan CF (1985) Hydrologic budgets for undisturbed and regenerating tropical rainforests on hillslopes in northeastern Costa Rica. In: Quinones E, Sanchez AV (eds) Proc Int Symp Tropical Hydrology and 2nd Caribbean Islands Water Resources Congress, San Juan, Puerto Rico. Am Water Resources Assoc, pp 11–15

Passioura JB (1988) Response to Dr. PJ Kramer's article "Changing concepts regarding plant water relations". Plant Cell Environ 11:569–571

Patric JH (1973) Deforestation effects on soil moisture, stream flow, and water balance in the central Appalachians. US Dep Agric For Serv Res Pap No NE-259, pp 1–12

Pearce AJ, Stewart MK, Sklash MG (1986) Storm runoff generation in humid headwater catchments. 1. Where does the water come from? Water Resour Res 22:1263–1272

Peck AJ, Williamson DR (1987) Effects of forest clearing on groundwater. J Hydrol 94:47–65

Penman HL (1948) Natural evaporation from open water, bare soil and grass. Proc R Soc Ser A 193:120–145

Pereira AR, De Camargo AP (1989) Analysis of the criticism of Thornthwaite's equation for estimating potential evapotranspiration. Agric For Meteorol 46:149–157

Pilgrim DH, Huff DD (1983) Suspended sediments in rapid subsurface stormflow on a large field plot. Earth Surf Processes Landforms 8:451–463

Ponce SL, Meiman JR (1983) Water yield augmentation through forest and range management – issues for the future. Water Resour Bull 19:415–419

Prebble RE, Stirk GB (1980) Throughfall and stemflow on silver ironbark (*Eucalyptus melanophloia*) trees. Aust J Ecol 5:419–427

Pyatt DG (1984) The effect of afforestation on the quantity of water runoff. Research Information Note 83, 84 SSN. Forest Commission Surrey, UK

Rakhmanov VV (1970) Dependence of stream flow upon the percentage of forest cover of catchments. FAO/USSR Int Symp Forest Influence and Watershed Management, Hydrometeorological Center USSR, pp 55–69

Reifsnyder WE (1989) A tale of ten fallacies: the skeptical enquirer's view of the carbon dioxide/climate controversy. Agric For Meteorol 47:349–371

Ricca VT, Simmons PW, McGuinness JL, Taiganides EP (1970) Influence of land use on runoff from agricultural watersheds. Trans ASAE 13:187–190

Robinson M (1986) Changes in catchment runoff following drainage and afforestation. J Hydrol 86:71–84

Robinson M (1989) Small catchment studies of man's impact on flood flows: agricultural drainage and plantation forestry. Int Assoc Hydrol Sci No 187

Rodriguez-Iturbe I (1986) Scale of fluctuation of rainfall models. Water Resour Res 22:15s–37s

Rothacher J (1970) Increases in water yield following clearcut logging in the Pacific northwest. Water Resour Res 6:653–658

Running SW, Coughlan JC (1988) A general model of forest ecosystem processes for regional applications. 1. Hydrologic balance, canopy gas exchange and primary production processes. Ecol Model 42:125–154

Ruprecht JK, Schofield NJ (1989) Analysis of stream flow generation following deforestation in southwest Western Australia. J Hydrol 105:1–17

Rutter AJ (1963) Studies on the water relations of *Pinus silvestris* in plantation conditions. I. Measurements of rainfall and interception. J Ecol 51:191–203

Rutter AJ, Kershaw KA, Robins PC, Morton AJ (1971/72) A predictive model of rainfall interception. Agric Meteorol 9:367–384

Rutter AJ, Morton AJ, Robins PC (1975) A predictive model of rainfall interception in forests. II. Generalisation of the model and comparison with observations in some coniferous and hardwood stands. J Appl Ecol 12:367–380

Salati E, Vose PB (1984) Amazon Basin: a system in equilibrium. Science 225:129–138

Salati E, Dall'Olio A, Matsui E, Gat JR (1979) Recycling of water in the Amazon Basin: An isotopic study. Water Resour Res 15:1250–1258

Sartz RS (1973) Effect of forest cover removal on depth of soil freezing and overland flow. Soil Sci Soc Amer Proc 37:774–777

Savill PS, Weatherup STC (1974) The effect of afforestation on the water runoff in the Woodburn catchment area. Forestry 47:45–56

Scheidegger AE (1968) Horton's law of stream numbers. Water Resour Res 18:877–886

Schmidt RA, Troendle CA (1989) Snowfall into a forest and clearing. J Hydrol 110:335–348

Schulze E-D, Čermák J, Matyssek R, Penka M, Zimmerman R, Vasicek F, Gries W, Kučera J (1985) Canopy transpiration and water fluxed in the xylem of the trunk of *Larix* and *Picea* trees – a comparison of xylem flow, porometer and cuvette measurements. Oecologia 66:475–483

Schulze E.-D, Steudle E, Gollan T, Schurr U (1988) Response to Dr PJ Kramer's article "changing concepts regarding plant water relations". Plant Cell Environ 11:573–576

Sharma ML (1984) Evapotranspiration from a eucalyptus community. Agric Water Manage 8:41–56

Sharma ML (ed) (1989) Groundwater recharge Proc Symp Groundwater Recharge, Mandurah, WA, 1987. A.A. Balkema, Rotterdam, 323 pp

Sharma ML, Craig AB (1989) Comparative recharge rates beneath *Banksia* woodland and two pine plantations on the Gnangara mound, Western Australia. In: Sharma ML (ed) Groundwater recharge. AA.Balkema Rotterdam, pp 171–184

Sharma ML, Johnston CD, Barron RJW (1982) Soil water and groundwater responses to forest clearing in a paired catchment study in south-western Australia. 1st Natl Symp Forest Hydrology. The Institution of Engineers, Australia, Publ No 82/6, pp 118–123

Sharma ML, Barron RJW, Williamson DR (1987a) Soil water dynamics of lateritic catchments as affected by forest clearing for pasture. J Hydrol 94:29–46

Sharma ML, Barron RJW, Fernie MS (1987b) Areal distribution of infiltration parameters and some soil physical properties in lateritic catchments. J Hydrol 94:109–127

Shuttleworth WJ (1977) The exchange of wind-driven fog and mist between vegetation and the atmosphere. Boundary-Layer Meteorol 12:463–489

Shuttleworth WJ (1979) Below-canopy fluxes in a simplified one-dimensional theoretical description of the vegetation-atmosphere interaction. Boundary Layer Meteorol 17:315–331

Shuttleworth WJ (1988a) Macrohydrology – the new challenge for process hydrology. J Hydrol 100:31–56

Shuttleworth WJ (1988b) Evaporation from Amazonian rainforest. Proc R Soc Lond B233:321–346

Sklash MG (1990) Environmental isotope studies of storm and snowmelt runoff generation. In: Anderson MG, Burt TP (eds) Process studies in hillslope hydrology. John Wiley, Chichester

Sklash MG, Stewart MK, Pearce AJ (1986) Storm runoff generation in humid headwater catchments. 2. A case study of hillslope and low-order stream response. Water Resour Res 22:1273–1282

Smith MK (1974) Throughfall, stream flow and intercpetion in pine and eucalypt forest. Aust For 36:190–197

Stewart JB (1977) Evaporation from the wet canopy of a pine forest. Water Resour Res 13:915–921

Stewart JB (1984) Measurement and prediction of evaporation from forested and agricultural catchments. Agric Water Manage 8:1–28

Stewart JB, Gay LW (1989) Preliminary modelling of transpiration from the FIFE site in Kansas. Agric For Meteorol 48:305–315

Stoertz MW, Bradbury KR (1989) Mapping recharge areas using a ground-water flow model. A case study. Groundwater 27:220–228

Stogsdill WR Jr, Wittwer RF, Hennessy TC, Dougherty PM (1989) Relationship between throughfall and stand density in a *Pinus taeda* plantation. For Ecol Manage 29:105–113

Stoneman GL, Schofield NJ (1989) Silviculture for water production in jarrah forest of Western Australia: an evaluation. For Ecol Manage 27:273–293

Stoneman GL, Rose PW, Borg H (1988) Recovery of forest density after logging in the southern forest of Western Australia. Tech Rept No 19, Dept. of Conservation and Land Management, Como, West Australia

Swank WT, Miner NH (1968) Conversion of hardwood-covered watersheds to white pine reduces water yield. Water Resour Res 4:947–954

Taniguchi M, Kayane I (1986) Changes in soil temperature caused by infiltration of snowmelt water. In Proc Budapest Symp, modelling snowmelt-induced processes. IAHS Publ No 155, pp 93–101

Trimble SW, Weirich FH (1987) Reforestation reduces stream flow in the southeastern United States. J Soil Water Conserv 42:274–276

Trimble SW, Weirich FH, Hoag BL (1987) Reforestation and the reduction of water yield on the southern Piedmont since circa 1940. Water Resour Res 23:425–437

Troendle, CA, King RM (1987) The effect of partial and clearcutting on stream flow at Deadhorse Creek, Colorado. J Hydrol 90:145–157

Turner KM (1984) Discussion "The potential for increasing stream flow from Sierra Nevada watersheds" by Kattelman RC et al. Water Resour Bull 20:453–454

Turner JV, Macpherson DK, Stokes RA (1987) The mechanisms of catchment flow processes using natural variations in deuterium and oxygen-18. J Hydrol 94:143–162

Turner NC (1986) Crop water deficits: a decade of progress. Adv Agron 39:1–51

Van Lill WS, Kruger FJ, Van Wyk DB (1980) The effect of afforestation with *Eucalyptus grandis* Hill ex Maiden and *Pinus patula* Schlect. et Cham on stream flow from experimental catchments at Mokobulaan, Transvaal. J Hydrol 48:107–118

Van Wyk DB (1987) Some effects of afforestation on stream flow in Western Cape Province, South Africa. Water SA 13:31–36

Van Zyl WH, De Jager JM (1987a) Accuracy of the Penman–Monteith equation adjusted for atmospheric stability. Agric For Meteorol 41:57–64

Van Zyl WH, De Jager JM (1987b) Estimating evapotranspiration from wheat using weather measurements and carborundum or Piche evaporimeter. Agric For Meteorol 41:65–75

Verry ES, Lewis JR, Brooks KN (1983) Aspen clearcutting increases snowmelt and storm flow peaks in north central Minnesota. Water Resour Bull 19:59–67

Walker J, Penridge LK (1987) FOL-PROF: a Fortran-77 package for the generation of foliage profiles. Part 1. User Manual. CSIRO Division of Water Resources Research Tech Memo 87/9. CSIRO Institute of Natural Resources and Environment, Canberra, pp 1–27

Walker J, Jupp DLB, Penridge LK, Tian G (1986) Interpretation of vegetation structure in Landsat MSS imagery: a case study in disturbed semi-arid woodlands. Part 1. Field data analysis. J Environ Manage 23:19–35

Wierda A, Veen AWL, Hutjes RWA (1989) Infiltration at the Tai rain forest (Ivory Coast): measurements and modelling. *Hydrol Processes* 3:371–382

Williamson DR, Stokes RA, Ruprecht JK (1987) Response of input and output of water and chloride to clearing for agriculture. J Hydrol 94:1–28

Wood EF (1983) Hydrology 1979–1982. US National Report to IUGG 1979–82. Eos Trans AGU 64:457

Subject Index

The manufacturer's authorised representative in the EU is Springer
Nature Customer Service Centre GmbH, Europaplatz 3, 69115 Heidelberg,
Germany. If you have any concerns regarding our products, please
contact ProductSafety@springernature.com

Printed and bound by CPI Group (UK) Ltd, Croydon, CR0 4YY
24/04/2026
02096348-0014